PYROGENIC MINERAL STABILITY, LOWER MEMBER OF THE JOHN DAY FORMATION, EASTERN OREGON

PYROGENIC MINERAL STABILITY, LOWER MEMBER OF THE JOHN DAY FORMATION, EASTERN OREGON

BY

RICHARD V. FISHER

UNIVERSITY OF CALIFORNIA PRESS
BERKELEY AND LOS ANGELES
1968

University of California Publications in Geological Sciences

Volume 75

Approved for publication September 15, 1967
Issued April 26, 1968
Price, $1.50

University of California Press
Berkeley and Los Angeles
California

Cambridge University Press
London, England

CONTENTS

PYROGENIC MINERAL STABILITY, LOWER MEMBER OF THE JOHN DAY FORMATION, EASTERN OREGON

BY

RICHARD V. FISHER

ABSTRACT

The stability sequence of pyrogenic minerals from altered dacitic volcaniclastic rocks in the lower member of the John Day formation (late Oligocene), eastern Oregon, varies with their different depositional environments. The original ash accumulated on the gently rolling erosion surface of a tropical or subtropical landmass. Ash that fell on well-drained upland areas altered to red rocks. Drab and mottled rocks developed in lowlands, and beneath oxidized soil zones in poorly drained areas. In places, drab, well-bedded, and fine-grained lacustrine tuffs interfinger with poorly bedded red and drab beds.

Comparison of Al_2O_3, TiO_2, and total iron in rocks from the John Day member with those in Hawaiian laterites indicates that the red John Day tuffs are immature laterites. Presence of kaolinite and red ferric oxides suggests, for the red rocks, an environment with relatively low pH and high Eh. Montmorillonite and some carbonaceous matter in the drab beds suggest an environment with an alkaline pH and a fairly low Eh. The lacustrine tuffs, with kaolin and usually abundant carbonaceous matter, were altered in an environment with relatively low pH and low Eh.

Pyrogenic minerals within many layers were probably subjected to more than one chemical environment, although their relative abundance and surface textures indicate the following general stability relations:

Red rocks (formed in acid, oxidizing environment)
zircon > magnetite > ilmenite > quartz > apatite > sodic plagioclase > potassic feldspar > calcic plagioclase

Drab rocks (formed in alkaline, reducing environment)
zircon > apatite > ilmenite > quartz > hornblende > magnetite > sodic plagioclase > potassic feldspar > calcic plagioclase > clinopyroxene

Stability relations of pyrogenic minerals in the fine-grained lacustrine tuffs are less clear than in the land-laid tuffs because of probable sorting factors in the ancient lakes. The presence of kaolin and the rarity of pyrogenic magnetite indicate deposition in an acid, reducing environment.

INTRODUCTION

General Statement

THIS REPORT concerns the stability of pyrogenic minerals in altered rocks of the lower member of the John Day formation (Oligo-Miocene), eastern Oregon. It supplements some of the work of Hay (1963) on diagenetic alteration of the formation.

Mapping of the John Day formation began in 1957, with subsequent field seasons during the summers of 1959–1964. Detailed mapping and measurement of twenty-two stratigraphic sections were done in the Kimberly, Picture Gorge, and Monument quadrangles; rapid reconnaissance was done in the adjacent Richmond, Spray, and Courtrock quadrangles (fig. 1). Laboratory investigation of rocks from the lower John Day member in the Picture Gorge quadrangle provides most of the data, and the basis, for interpretations herein.

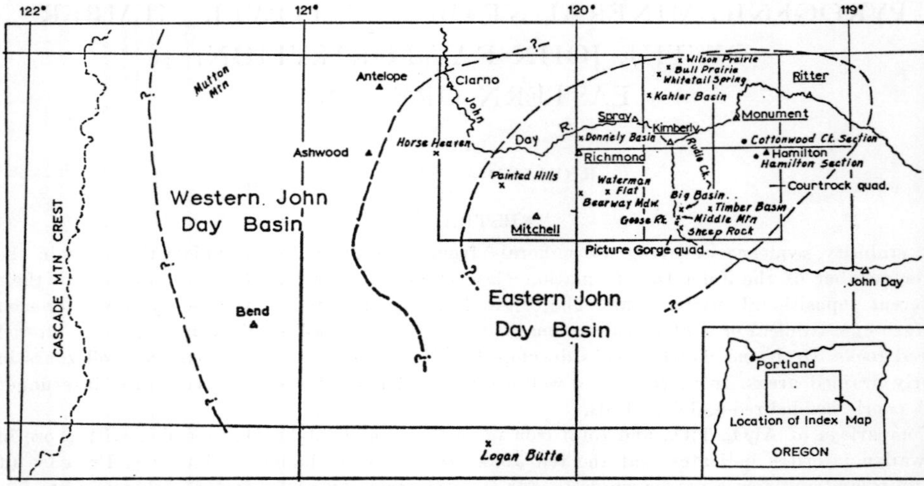

Fig. 1. John Day Basin area. Underlined names indicate quadrangles. Dashed lines show inferred limits of western and eastern John Day Basins.

Laboratory Procedure

Friable tuff samples were first coarsely broken in a mortar and screened through a 1-mm screen to eliminate small, crushed fragments. The walnut- to pea-size fragments were then rolled for 2 hours in jars containing a steel roller, water, and a dispersing agent. The sediment was then washed through a $\frac{1}{16}$-mm screen, and the coarse fraction was scrubbed for 2 hours in a sonic vibrator; longer periods in the sonic vibrator caused abrasion-frosting of the grains. The cleaned sand was then dried and placed in bromoform to separate the heavy minerals (specific gravity (> 2.89) from the light fraction. Magnetite was removed from the nonmagnetic heavy minerals by repeatedly passing a small Alnico hand magnet about $\frac{1}{4}$ inch above the sample. The light-mineral fraction was run through a Franz magnetic separator set at 1.5 amperes with a 5-degree side tilt to obtain the nonmagnetic minerals. The nonmagnetic fraction consisted of feldspar and zeolite; they were separated by a mixture of bromoform and dimethyl sulphoxide (specific gravity $= 2.40$). Thus three sample splits were obtained: feldspar, magnetite, and nonmagnetic heavy minerals. The nonmagnetic heavies were mounted in balsam, and 300 grains on each slide were identified. The feldspar separates were screened through sieves to determine size parameters. Surface textures of magnetite and feldspars were examined in reflected light with a Zeiss stereoscope. Compositions of feldspars were determined by index of refraction measurements, using the curves of Smith (1958).

Acknowledgments

This paper is an outgrowth of work on the John Day formation under National Science Foundation Grants G–18642 and GP–2148. Edward Oliver, Harold Mattraw, and Casey Moore helped greatly in the laboratory. Lloyd Edwards made some partial chemical analyses of the lower John Day siltstones. Discussions of problems with Richard L. Hay, Aaron C. Waters, and Ray E. Wilcox are acknowledged. Richard Migues was my field assistant during the summer of 1962.

The manuscript, illustrations, and analyses were completed while I was associated with the Hawaii Institute of Geophysics. This paper is Contribution No. 138 of the Hawaii Institute of Geophysics.

STRATIGRAPHY

General Statement

Rocks of the mapped area range in age from Paleozoic to Pliocene. From oldest to youngest, they include pre-Cretaceous metamorphic rocks, Cretaceous conglomerates and sandstones, Eocene volcanic rocks of the Clarno formation, the Oligo-Miocene volcaniclastic John Day formation, and the Picture Gorge basalt (Waters, 1961). Pleistocene and Recent deposits include loessial soils above unweathered basalt flows of the upland plateaus, and alluvial and landslide deposits on slopes and valley margins. White, volcanic ash derived from Crater Lake (Mount Mazama) (Roy E. Wilcox, personal communication) is interbedded, in places, with old alluvium, but it has generally been reworked into wedge-shaped deposits at the confluence of streams.

Pre-Cretaceous Rocks

Small areas of pre-Cretaceous crystalline rocks are exposed in the Picture Gorge quadrangle north and south of the Middle Mountain fault, in small slices in the Middle Mountain fault zone, and in Timber Basin east of Middle Mountain (fig. 1). These rocks include fine-grained marble, highly sheared quartzite, and chlorite-muscovite schist. They crop out beneath Cretaceous rocks in secs. 27 and 28, T. 11 S., R. 26 E., east of Middle Mountain, and directly underlie John Day rocks in secs. 17 and 18, T. 11 S., R. 26E.

Cretaceous Rocks

An unnamed sequence of Cretaceous sedimentary rocks, unconformable above the crystalline rocks, crops out in several small areas in Big Basin, and also on the south side of Middle Mountain (Fisher, 1964) and in Timber Basin. Cretaceous rocks are also exposed in Waterman Flats and Bearway Meadow in the Richmond quadrangle (fig. 1), about 25 miles west of Big Basin, where they are overlain by a thick sequence of Clarno volcanic rocks. In the Big Basin area, Cretaceous rocks are overlain by the John Day formation.

In the Big Basin area, the rocks are predominantly thick-bedded conglomerate layers with interbedded sandstone and some shale; they vary from 0 to 500 feet in thickness. In Bearway Meadow, conglomerate and sandstone similar to those in Big Basin overlie dark fissile and platy shale, a sequence correlative with conglomerate and shale farther west in the Mitchell quadrangle dated as Albian (Jones, 1960; Packard, 1956).

Fragments in the conglomerate beds are well-rounded pebbles and cobbles of quartzite, metavolcanic rocks, and chert; sand-sized clasts include subrounded to subangular quartz, orthoclase, plagioclase, and conspicuous accessory biotite.

Clarno Formation

Andesite and basalt of the Clarno formation occur in the Picture Gorge quadrangle in a small area near Goose Rock (fig. 1) on the west side of the John Day River near Oregon State Highway 19, and near Rudio Creek in the northeastern

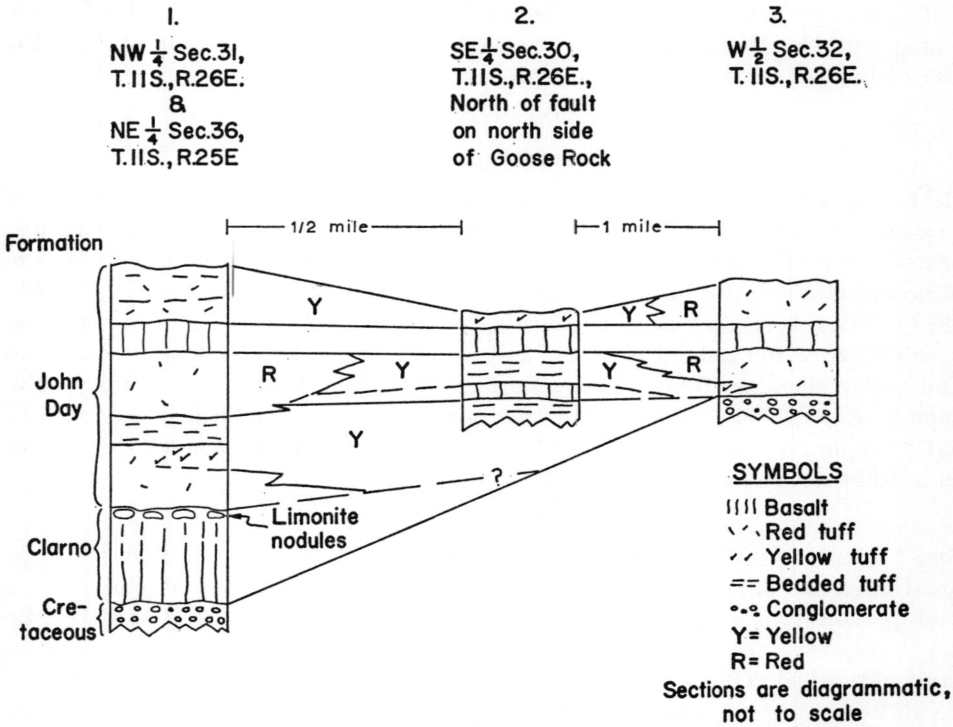

Fig. 2. Stratigraphic relationships between the lower member of the John Day formation, Clarno volcanic rocks, and Cretaceous rocks near Goose Rock, Picture Gorge quadrangle.

part of the quadrangle. More extensive exposures of Clarno rocks crop out along the northern edge of the Kimberly quadrangle where they are overlain by thin, discontinuous patches of John Day rocks and a thin sequence of Columbia River basalt flows. Extensive exposures of Clarno andesite and basalt also occur in the central parts of the Richmond and Courtrock quadrangles.

Post–Clarno, Pre–John Day Soil Zone

A red, clay-rich soil zone blankets the uppermost part of the Clarno volcanic sequence, or else lies on older formations where Clarno rocks are absent. As first described by Waters *et al.* (1951) in the Horse Heaven mining district (fig. 1), the zone mantles a gently rolling surface eroded on Clarno rocks, averages between 10 and 20 feet and ranges from a few inches to as much as 100 feet or more in thickness, and is overlain by basaltic andesite or rhyolite.

This ancient soil occurs over a wide area in eastern Oregon. It has been described a few miles west of the Horse Heaven mining district (Peck, 1964); in the Mitchell quadrangle east of the mining district (Hay, 1962): and in the Picture Gorge quadrangle (Fisher, 1964). In Rudio Creek, the soil zone lies on Clarno rocks, but near Middle Mountain it occurs as a hardpan on tilted Cretaceous conglomerate and sandstone (Fisher, 1964). Clarno basalt near Goose Rock is deeply weathered to a red lateritic soil which contains hollow limonite concretions up to 1 foot in diameter.

The contact between John Day and Clarno rocks is often difficult to place. In the Horse Heaven mining district, Waters *et al.* (1951) use the red-soil zone to separate the formations, a practice followed here.

The complexity of this boundary is illustrated by relationships in the Picture Gorge quadrangle near Middle Mountain (fig. 2). The weathered surface is developed on Cretaceous rocks except at one locality near Goose Rock, where the Cretaceous rocks are directly overlain by basalt with a deeply weathered top containing limonite concretions (see fig. 2, sec. 1). Similar olivine basalt flows east of Goose Rock are assigned to the John Day formation because they are underlain by lacustrine tuff containing leaves equivalent to the Cant Ranch and Bridge Creek flora of John Day age (Chaney, 1927) (fig. 2, sec. 2). Farther east, the weathered surface is developed on Cretaceous rocks and is overlain by red rocks of the lower John Day member (fig. 2, sec. 3).

Olivine basalt flows associated with red beds and lacustrine tuffs also occur in Donnely Basin in the Spray quadrangle (Lindsley, 1960; Hay, 1963), and near Painted Hills (Hay, 1962, 1963). Hay (1963, p. 205) reports eight flows in the Donnely Basin area which total 400 feet in thickness. Lindsley (1960) assigns the lower seven flows to the Clarno formation, whereas Hay (1963) assigns them to the John Day formation. A regional petrographic study of the John Day and upper Clarno basalt flows is needed to resolve this problem, but is not attempted here.

John Day Formation

The John Day formation, consisting of salic volcaniclastic rocks, occurs within two major basins of deposition (fig. 1) separated by a pre–John Day structural high (Fisher, 1967). The western basin which lies adjacent to the Cascade Mountains includes abundant coarse-grained volcaniclastics and salic plugs (Waters *et al.*, 1951; Peck, 1964). John Day rocks in the eastern basin are predominantly massive, fine- to medium-grained volcaniclastic rocks interbedded with some coarse-grained air-fall tuffs. Nine or more ignimbrite units have been reported in the western basin (Peck, 1964), and one layer with two cooling units occurs in the eastern basin (Hay, 1963; Fisher, 1966). Unless otherwise noted, "John Day formation" applies here only to rocks of the eastern basin where Merriam (1901) originally subdivided the John Day sequence into three "divisions," primarily on the basis of color: a lower division composed of red volcanic claystones and siltstones with some interbedded white or buff tuff layers; a middle division of drab to bluish-green beds; and an upper division composed of buff ashy deposits, with sand and gravel at the top in places.

Detailed mapping and section measurement in the Kimberly and Picture Gorge quadrangles show that Merriam's upper division is divisible into two parts. The "sands and gravels" near the top of Merriam's upper division disconformably overlap the lower part of his upper division as well as older John Day rocks south of the town of Kimberly (Fisher, 1967).

In the present report I deal only with the lowest member of the John Day formation—Merriam's lower division.

LOWER MEMBER, JOHN DAY FORMATION

General Description

Red claystones and siltstones of the lower John Day member occur as far west as Mutton Mountains (T. 7 and 8 S., R. 13 E., AMS, Bend Sheet), as far east as Hamilton (T. 10 S., R. 28 E., AMS, Canyon City Sheet), and as far south as Logan Butte (T. 19 S., R. 20 E., AMS, Crescent Sheet) (fig. 1). Poorly bedded red claystones and siltstones are the most abundant rock type, but gray, brown, and buff rocks are common in many places. Some areas include interbedded white air-fall tuff layers in marked contrast with the brilliant red rocks. Well-bedded, buff, leaf-bearing lacustrine tuffs and olivine basalt flows crop out in local areas near the base of the member (Hay, 1962; Fisher, 1964).

The lower member, deposited over an uneven erosion surface, is variable in thickness. Hay (1962) describes 450 feet of red and yellow claystones and bedded lacustrine tuffs in the Painted Hills area (Mitchell quadrangle), 34 miles west of Big Basin (Picture Gorge quadrangle). Near Hamilton, 16 miles northeast of Big Basin, Fisher and Wilcox (1960) measured 240 feet of red claystones and siltstones with three interbedded white air-fall tuff units. A. C. Waters (personal communication) reports that the lower member is 500 to 600 feet thick in areas near the Cascade Mountains. Red, and gray to yellow, volcanic claystones and siltstones, totaling 150 feet in thickness, were measured in Big Basin.

In the following discussion I briefly describe the lower John Day member from a few areas within the investigated region, and then give a more detailed description of a measured section in the Big Basin area.

Monument–Hamilton Region

The lower member of the John Day formation is extensively exposed in the north-central part of the Courtrock quadrangle, and in a few places in the southern part of the Monument quadrangle. It has been examined near Hamilton in sec. 2, T. 10 S., R. 28 E. (Courtrock quadrangle) and in NE ¼ sec. 30, T. 9 S., R. 28 E. (Monument quadrangle) (Fisher and Wilcox, 1960) (fig. 1).

HAMILTON SECTION

The Hamilton section, well exposed along the eastern wall of a landslide area, consists of 240 feet of deep-red volcanic claystones and siltstones with three 1- to 5-foot thick, light-gray to pink air-fall tuff layers. The red rocks rest upon an intensely weathered surface of Clarno andesite surface with about 15 to 20 feet of relief. The top of the lower member is a 2- to 3-foot tuff bed containing abundant zeolitized shards and large, green, phyllite fragments. Coarse-grained, pale-red to gray tuffs of the middle John Day member overlie it.

The contact between the brownish-red weathered Clarno andesite and the deep-red John Day claystones is located by pyrogenic feldspars which occur in the John Day but not in the weathered Clarno.

COTTONWOOD CREEK SECTION

The phyllite-bearing air-fall tuff found in the Hamilton section also occurs in the upper 70 feet of the lower John Day member exposed in NE ¼ sec. 30, T. 9 S.,

R. 28 E. (Monument quadrangle). Unlike most of the lower John Day member, the rocks are pale-brown siltstones which are cut by a few thin silica veins.

Kahler Basin Area

The lower John Day member is thin, and crops out in only a few areas near the northern margin of the Kimberly quadrangle (fig. 1). Red claystones and siltstones, only 70 to 80 feet thick, rest on weathered olivine basalt of probable Clarno age in Kahler Basin (T. 7 S., R. 25 E.). On Wilson Prairie, about 8 miles northeast of Kahler Basin, outcrops of weathered Clarno andesite veneered with probable lower John Day member rocks underlie Yakima basalt (A. C. Waters, personal communication). A thin sequence of lower John Day member rocks also crops out near Whitetail Spring (sec. 12, T. 7 S., R. 25 E.) and in Bull Prairie (sec. 7, T. 7 S., R. 26 E.) (fig. 1).

Big Basin Area

INTRODUCTION

Rocks of the lower John Day member in Big Basin appear in several erosional windows cut through extensive landslide deposits; south of Middle Mountain they crop out east and west of Goose Rock and near Sheep Rock. They rest unconformably on Cretaceous conglomerate and sandstone except in SE ¼ sec. 18, T. 11 S., R. 26 E., where they lie on pre-Cretaceous metamorphic rocks.

Red claystones and siltstones with interbeds of yellowish-gray claystone and siltstone predominate. Well-bedded, buff to gray, fine-grained lacustrine tuffs and basalt flows crop out near Goose Rock, and in SE ¼ sec. 18, T. 11 S., R. 26 E. The dominantly red rocks occur on ancient hilltops carved on Cretaceous rocks, whereas the lake beds and basalt flows occur in ancient low areas (Fisher, 1964).

NONRED VOLCANICLASTIC ROCKS

Three kinds of nonred lower-member John Day volcaniclastic rocks occur in the Big Basin area, as well as elsewhere (Hay, 1962): (1) 3- to 4-foot-thick, gray to white air-fall layers; (2) drab gray to yellow-gray massive volcanic claystones and siltstones; and (3) yellow and gray leaf-bearing, laminated lacustrine tuff beds containing freshwater pelecypods and gastropods.

Light-gray air-fall tuff layers are interbedded with red claystone and siltstone in E. ½ sec. 6, T. 12 S., R. 26 E., near Sheep Rock. They also occur in the Hamilton area.

The yellowish-gray, massive rocks occur in the upper part of the member in Big Basin, and near the base of the member near Sheep Rock. East and west of Goose Rock these drab-colored rocks interfinger with red claystones (fig. 2), as in Painted Hills State Park, 34 miles west of Big Basin (Hay, 1962).

Laminated yellow to gray lacustrine tuffs in the Big Basin area occur in SE ¼ sec. 18, SE ¼ sec. 30, and in the center of sec. 32, T. 11 S., R. 26 E. Numerous gypsum veinlets up to 1 cm wide cut the beds, and in some places the rocks are colored bright yellow by natrojarosite which occurs within fractures, as a cement, and in veinlets mixed with gypsum. Some natrojarosite-bearing rocks are silicified.

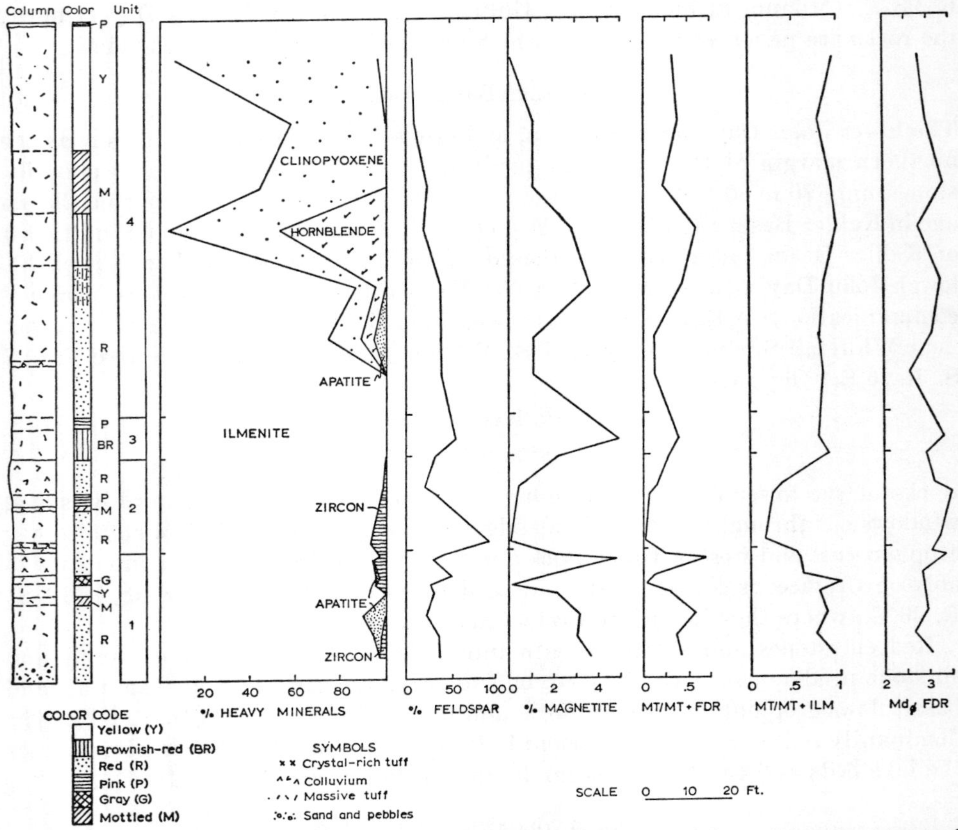

Fig. 3. Graphic illustration of vertical variations of heavy minerals, weight percentage of feldspar and magnetite, magnetite-feldspar and magnetite-ilmenite ratios, and median diameter (Mdφ) of feldspar samples.

MEASURED SECTION

A 150-foot section of red and drab rocks of the lower John Day member in Big Basin (SE ¼ sec. 17, T. 11 S., R. 26 E., Picture Gorge quadrangle) is divided into four units based on soil structures and color (fig. 3; table 1, in pocket). The units cannot be traced to other areas and therefore are not stratigraphic markers. Claystones and siltstones, which grade from red at the base to brown and yellow at the top, and are somewhat coarser grained at the top, make up the section. Landslide rubble composed of Picture Gorge basalt and John Day tuff fragments overlies the measured section.

Unit 1, the basal unit (see table 1, in pocket), is dominantly red claystone which grades upward to red-yellow mottled rocks and gray claystone which underlie a red-brown soil horizon. Near the base of the unit are pebbles and minerals derived from the underlying Cretaceous rocks. Unit 2, above unit 1, grades from dark-red rocks upward to a pale-red layer which underlies a relatively resistant red-soil horizon. The soil horizon contains yellowish-brown pisolitic areas. Unit 3, which grades upward from dark- to light-red rocks, is overlain by dark-red claystones of unit 4. The division between units 3 and 4 is arbitrarily placed at the transitional color change from light- to dark-red rocks. Unit 4 consists of more than half of the measured section; its red basal rocks grade upward to gray and brown rocks at the top.

Petrology and Mineralogy

INTRODUCTION

Rocks in the lower member of the John Day formation range from claystone to lapilli tuff. Altered glass shards, pumice fragments, and a few pyrogenic minerals are their main constituents. Relict shard and pumice textures are preserved in most rocks.

Alteration products, chiefly produced from volcanic glass, including montmorillonite, heulandite, clinoptilolite, kaolinite, and goethite in the red rocks. In the drab rocks, alteration products are mainly montmorillonite and clinoptilolite; and in the lacustrine beds they include montmorillonite, kaolinite, and quartz. Gypsum occurs as discrete crystals and in veins within the poorly bedded drab rocks. Gypsum veins associated with natrojarosite occur in places within the lacustrine tuff. Within vesicles of altered basalt are found chlorite, calcite, chalcedony, and celadonite.

Pyrogenic minerals in the Big Basin section consist of andesine and lesser amounts of oligoclase and labradorite; some rocks contain small amounts of potassium feldspar, and nearly all contain small quantities of quartz. The dominant accessory mineral is magnetite, followed in abundance by ilmenite. Zircon, apatite, hornblende, oxyhornblende, and clinopyroxenes occur in small amounts in many rocks. An unusual mineral assemblage within the phyllite-bearing tuff (Monument–Hamilton area) includes relatively abundant quartz, sanidine, biotite, and rare magnetite.

COTTONWOOD CREEK SECTION

Lower-member John Day rocks in the Cottonwood Creek section contain abundant lithic fragments. In some sandy layers, they form as much as 35 percent of the sand-size material, and are as much as 2mm in longest dimension. The lithic fragments are composed of angular to subangular rhyolite, dacite, andesite, pumice, and dark-gray to red scoria. The pumice fragments are up to 8mm in longest dimension. Phyllite fragments up to 4cm occur in the uppermost layer of the section.

Pyrogenic minerals include oligoclase and andesine, and accessory magnetite and ilmenite. Less abundant are potassium feldspar, quartz, zircon, and apatite. Feldspars are chiefly angular to subangular anhedral grains; euhedral crystals are rare. Some were highly rounded by resorption before extrusion. Median diameters of feldspar samples vary from 2.39 ϕ* (0.19mm) to 3.07 ϕ (0.12mm), and average 2.86 ϕ (0.16mm); the largest observed crystal was an unbroken, water-clear andesine grain 1.4mm long. Potassium feldspar grains are milky in appearance and are more corroded than the plagioclase. Quartz crystals are commonly doubly terminated, unpitted, glassy, and unstained. Etched and frosted magnetite

* Phi (ϕ) units, proposed by Krumbein (1934), are Wentworth size limits expressed as powers of 2; $\text{phi} = -\log_2 \text{diam (mm)}$.

occurs as unbroken and broken euhedral crystals, most with rounded terminations. A few are smooth, shiny crystals, and some are encased by a thin jacket of altered glass showing bubble-wall texture (Fisher, 1963). Ilmenite grains are euhedral to anhedral with smooth, unpitted surfaces and splendent luster; most have sharp to rounded terminations. Many are incompletely formed crystals with pits containing altered glass, and some have bubble-wall texture. Small, euhedral zircon crystals occur in some rocks but are rare.

Dominant feldspar of the uppermost phyllite-bearing tuff is sodic plagioclase, but potassic feldspar is more abundant than in most of the lower-member rocks; the sodic plagioclase is more highly pitted than the potassic feldspar. Slightly pitted quartz is glassy, unstained, and usually doubly terminated. Ilmenite, which forms about a third of the nonmagnetic mineral separates, is in black, shiny, unpitted, euhedral crystals. Other minerals include unaltered euhedral biotite, rare magnetite euhedra unaltered or slightly pitted, and zircon euhedra which are colorless in reflected light.

Lower-member John Day rocks of the Monument–Hamilton region are notably coarser grained than those in Big Basin, suggesting that one source of John Day ash may have been to the north. The source of the topmost phyllite-bearing tuff is not known.

BIG BASIN AREA

Air-Fall Tuff

Light-gray air-fall tuff interbedded with red siltstones near Sheep Rock (E ½ sec. 6, T. 12 S., R. 26 E.) contains zeolitized shards and pyrogenic minerals set in a pasty, cream-colored montmorillonite matrix. Pyrogenic minerals make up less than 5 percent of the sand-size fraction of most rocks, and one specimen contained less than 0.10 percent.

Oligoclase and andesine are angular to subangular, and usually pitted, but are glassy with little internal alteration. Potassium feldspar occurs in some beds. Rare quartz grains are angular and chemically pitted; most are stained red. Heavy minerals include corroded pale-pink zircon with internal sinuous cavities and abundant unidentified inclusions; stubby apatite crystals with rounded terminations; splendent, smooth to moderately pitted black ilmenite grains with sharp to rounded terminations (ilmenite is the most common heavy mineral of the air-fall tuffs); rare pale-green clinopyroxene crystals with sawtooth terminations; dull and pitted anhedral to euhedral magnetite crystals in trace amounts. Minor quantities of small, angular to subangular, pink to gray felsitic fragments occur in most of the light-colored air-fall tuffs.

Lacustrine Tuff

The lacustrine tuff, in thinly bedded layers, contains fine-grained clastic particles which include sand-size pumice fragments, and sparse pyrogenic grains. Gypsum veinlets and, in some places, bright-yellow natrojarosite are present. Diagenetic kaolinite, associated with montmorillonite, is more abundant than in other lower-member rocks, and is associated with microcrystalline quartz.

Feldspars are predominantly water-clear oligoclase and sodic andesine, but range from potassic feldspar to labradorite. They are generally subangular, irregular, unpitted grains which show more of a cloudy internal alteration than is present in feldspars from the air-fall tuffs. Potassic feldspar shows more chemical pitting, is more ragged in outline, and has more internal clay-altered patches than the plagioclase grains. Pitted, irregular magnetite grains are rare or absent. Ilmenite, far more abundant than magnetite, is black and unpitted, and is in the form of sharply euhedral crystals. Pale-pink, sharply euhedral zircon, and stubby, slightly rounded apatite crystals, occur in trace amounts. A few grains of etched

TABLE 2
CALCULATED MINERAL RATIOS, BIG BASIN MEASURED SECTION,[a] PICTURE GORGE QUADRANGLE

Sample number	Mt/Mt + Fdr[b]	Mt/Mt + Ilm[b]	Recalculated to 100 percent		
			Nonmagnetic heavies	Magnetite	Feldspar (+ Qtz)
62F388	.086	.935	0.6	8.4	90.6
389	.079	.749	0.3	7.9	89.2
390	.117	.860	2.3	11.6	86.0
391	.061	.786	1.5	6.1	92.5
392	.004	.999	0.1	0.3	99.6
393	.029	.599	2.2	2.9	95.0
394	.139	.596	10.0	12.5	77.5
395	.043	.538	12.7	3.7	83.5
396	.001	.183	0.6	0.2	99.2
397	.014	.272	2.6	1.3	96.1
398	.020	.386	5.1	2.1	93.0
399	.059	.890	4.7	5.6	89.5
400	.079	.775	2.7	7.7	89.7
401	.028	.815	2.8	2.8	94.5
402	.028	.785	3.5	2.7	93.8
404	.084	.840	1.2	8.3	90.5
405	.199	.978	10.0	10.6	79.5
406	.041	.915	0.8	3.8	95.5
407	.079	.752	4.4	9.6	86.0
408	.081	.947	7.4	7.4	85.4

[a] SE ¼ Sec. 17, T. 11 S., R. 26 E.
[b] Mt = magnetite; Fdr = feldspar; Ilm = ilmenite.

clinopyroxene with sawtooth ends, and brown to green unaltered hornblende, occur in leaf-bearing lacustrine tuffs in NE ¼ sec. 36, T. 11 S., R. 25 E., near Goose Rock.

Basalt

Basalt flows in the lower John Day member, with ophitic or pilotaxitic texture, contain olivine, clinopyroxene, and labradorite. One flow contains abundant long, highly twinned labradorite laths, and a purple-tinted clinopyroxene ("titanaugite") which is weakly pleochroic and has a $+2V = 45°$. Greenish-brown fibrous material completely or partly replaces olivine.

Measured Section

Introduction.—Pyrogenic minerals from the poorly bedded, red, drab, and mottled claystones and siltstones of the Big Basin measured section were studied in more detail than those in the other sections. Size data, percentages, and descriptions of feldspar, quartz, magnetite, and nonmagnetic heavies are shown in table 1; mineral ratios are shown in table 2. Figure 3 graphically depicts vertical variations of the minerals, and summarizes most of the data given in tables 1 (in pocket) and 2.

Pyrogenic minerals, general description.—Plagioclase varies from sharp, unpitted euhedra to etched, angular grains with abundant external pits. Oligoclase and andesine are the most abundant feldspars, but some anorthoclase and labradorite grains occur. Most calcic plagioclase is completely altered to clays. The different alteration rate of plagioclases of different composition occurring in the same rock is illustrated in plate 1 (figs. 1, 2). Oligoclase rimmed with montmorillonite (pl. 1, fig. 1) has relatively smooth borders, but further alteration may have been inhibited by blanketing of the clay. Some crystals, however, may be coated with montmorillonite altered from original glass. During disaggregation, the clay is usually stripped off, leaving grains with smooth, unaltered faces. Calcic andesine (pl. 1, fig. 2) has highly crenulated borders and internal darkened areas. Grains with crenulated borders remain reddish after disaggregation because of adhering clay in the reentrant areas.

Quartz is not abundant, but may constitute as much as 10 percent by volume of the feldspar. Most quartz grains are angular, anhedral fragments, and are stained red on etched and frosted surfaces. A few highly frosted euhedral dipyramids occur in most samples.

Magnetite is in black, splendent grains which vary from euhedral octahedrons to nearly perfect spheres. Euhedral grains have sharp to rounded terminations, and are shiny even if pitted. Anhedral grains are highly pitted, frosted, and of dull luster. Many magnetite crystals are coated with altered glass, proving that they are pyrogenic.

Ilmenite forms black, splendent crystals, although some are pitted and dull. They vary from sharp euhedra to smooth spheres; broken euhedra occur in most samples. Some crystals have structures that resemble zoning, but are formed by progressive growth around rounded cores that outwardly become more euhedral.

Clinopyroxenes are pale- to yellow-green crystals; most are highly corroded and have sawtooth terminations. Etched hypersthene is very rare.

Small amounts of dark-brown oxyhornblende and dark–greenish-brown hornblende occur as fragments which are not etched or frosted.

Sparse zircon euhedra are pale or medium pink; length:breadth ratios range from 1.5:1 to 5:1 and average about 2:1; the longest dimension is usually less than 0.1mm.

Stubby, colorless, apatite crystals commonly have slightly rounded terminations, but some have sharp terminations. They average about 0.2mm in length, and contain tiny rods and needle-shaped inclusions.

Pyrogenic mineral alteration and rock color.—There are notable as well as subtle differences in alteration features of pyrogenic minerals between the dark-red and the drab-colored to pale-red rocks of the measured section.

Pyrogenic grains of the red rocks are (1) sharply euhedral to rounded, and splendent but somewhat pitted magnetite grains; (2) euhedral, splendent, and smooth to pitted ilmenite crystals; (3) sharply euhedral pink to colorless zircon crystals; (4) strongly etched apatite grains with weak sawtooth terminations; and (5) frosted, reddish-stained, angular quartz grains, present only in small amounts. Oligoclase crystals are water clear and angular to rounded anhedra; they weather to irregular shapes, but are unpitted. Andesine is etched to highly ragged shapes with altered and cloudy red-stained interiors. Calcic andesine and labradorite crystals are altered to clays in most of the dark-red rocks. Potassic feldspars, present in some red rocks, are slightly more etched than plagioclase. All minerals are more highly altered in the red colluvial soil zones than in the massive red rocks.

Pyrogenic minerals in the drab, pale-red, and mottled rocks in the upper half of the measured section include magnetite, ilmenite, clinopyroxene, hornblende, apatite, zircon, feldspars, and quartz. Clinopyroxenes and hornblende are absent in the lower part of the section. As compared with the darker-red rocks, magnetite is more highly etched and pitted and less abundant except in the interlayered red-soil zones; ilmenite is less altered; apatite is more abundant and less etched; and most plagioclase shows more etching and more internal alteration to clay. As in the dark-red rocks, potassic feldspar is somewhat more etched than sodic feldspar. Quartz is frosted and angular, and is stained pink; zircon is pink to colorless and is sharply euhedral. Relics of iron-rich clinopyroxenes are altered to reddish-brown clay; pale-green clinopyroxenes are etched to sawtooth terminations. Hornblende grains are angular and anhedral, but usually are unaltered.

In the mottled red-drab rocks, thin sections show that, in the same specimen, feldspars are more highly altered in drab areas than in red areas. Mineral separations from different-colored areas of single mottled specimens were not made, but it is probable that differences in alteration of the other pyrogenic grains parallel those found between the more evenly colored drab and red rocks.

Vertical variations of pyrogenic minerals.—Sand-size fragments vary from about 8 to 20 percent by weight. In most samples, pyrogenic minerals make up less than half the sand; the rest is composed of zeolitized shards, altered pumice, and a few lithic fragments. Most shards and pumice were altered to soft clays and zeolites; otherwise the percentage of sand-size grains would be larger. Because textures of the altered volcanic siltstones and claystones cannot be characterized by size-frequency curves of whole-rock samples, only size parameters of feldspar grains were used.

Investigated mineral parameters are: relative percentages and kinds of heavy minerals; amount of feldspar calculated as weight percent of the total sand-size fraction; median diameter of the feldspar samples; amount of magnetite (weight percent of sand-size sample); the ratios, Mt/(Mt + Ilm) (magnetite-ilmenite ratio) and Mt/(Mt + Fdr) (magnetite-feldspar ratio).

Overall upward changes of investigated parameters are shown in figure 3. Important changes are that ferromagnesians are absent in the lower part of the section; feldspar and magnetite show a tendency to increase; the magnetite-ilmenite ratio tends to increase above unit 1; and feldspar median diameters show no trend. The apparent upward decrease in amount of feldspar and magnetite is owing to an increase in relative percentage of sand-size shards and pumice fragments, as indicated by the lack of trend in their ratio.

Within unit 1, the magnetite-ilmenite ratio decreases upward to its lowest

TABLE 3

ALUMINUM, IRON, AND TITANIUM CONTENTS OF LOWER-MEMBER JOHN DAY AND HAWAIIAN ROCKS AND SOILS

(In percentages)

Sample number	Al_2O_3	$(Fe_2O_3 + FeO)$	TiO_2	$Al_2O_3 \div (Al_2O_3 + \overline{Fe})$	Comment
1......	16.47	2.13	0.13	0.89	John Day, glass shards[a]
2......	11.60	1.75	0.20	0.87	John Day, glass shards[b]
3......	11.40	1.76	0.14	0.87	John Day, glass shards[b]
4......	11.70	1.68	0.12	0.88	John Day, glass shards[b]
5......	17.90	8.90	1.20	0.67	John Day, brown, lower member[b]
6......	17.20	9.21	0.94	0.65	John Day, red, lower member[c]
7......	5.24	14.69	0.96	0.26	John Day, gray, lower member[c]
8......	8.76	14.61	1.18	0.37	John Day, mottled, lower member[c]
9......	12.93	11.43	2.52	0.53	Av. Mauna Loa basalt, (Macdonald and Katsura, 1964)
10......	13.25	11.63	2.86	0.53	Av. Kilauea basalt (Macdonald and Katsura, 1964)
11, A_1[d]..	19.60	33.60	8.00	0.37	Soil profile I, 25–45 inches,[e] Lanai, Hawaii (Sherman *et al.*, 1948)
12, A_2[d]..	16.20	45.00	10.80	0.26	
13, A_1...	18.00	39.42	8.61	0.31	Soil profile II, 25–45 inches,[e] Kauai, Hawaii (Sherman *et al.*, 1948)
14, A_2...	7.81	58.02	14.44	0.12	
15, A_1...	13.20	43.24	17.36	0.23	Soil profile III, 60–150 inches,[e] Maui, Hawaii (Sherman *et al.*, 1948)
16, A_2...	13.20	48.08	18.92	0.22	
17, A_1...	15.80	33.52	7.68	0.32	Soil profile IV, 35–80 inches,[e] Maui, Hawaii (Sherman *et al.*, 1948)
18, A_2...	9.60	48.72	10.68	0.16	
19, A_1...	10.65	38.20	19.37	0.22	Soil profile V, 35–80 inches,[e] Kaui, Hawaii (Sherman *et al.*, 1948)
20, A_2...	17.85	48.50	24.69	0.27	
21, A_1...	18.80	36.20	10.40	0.34	Soil profile VI, 24–45 inches,[e] Oahu, Hawaii (Sherman *et al.*, 1948)
22, A_2...	25.80	27.20	6.40	0.49	

[a] Analyst: Hiroshi Asari, Japan Analytical Chemistry Research Institute, Tokyo, Japan.

[b] Analysts: P. Elmore, S. Botts, G. Chloe, and H. Smith, U.S. Geological Survey, using methods described in U.S. Geological Survey Bulletin 1144-A. For sample number 2, field number 61F46-G, laboratory number 159943; for sample number 3, field number 59F49-G, laboratory number 159945; for sample number 4, field number 59F57-G, laboratory number 159942; for sample number 5, field number 59F135, laboratory number 159941.

[c] Analyst: Lloyd Edwards, University of California, Santa Barbara. Partial analyses, including TiO_2, P_2O_5, MnO, total iron, and Al_2O_3.

[d] A_1 and A_2 refer to soil zones.

[e] Yearly rainfall average.

value in the capping colluvial soil zone, and is at a maximum in the drab claystones which contain a small absolute amount of the two oxides. The magnetite-feldspar ratio is high within the upper colluvial soil zone where both minerals are highly etched and the feldspars have a relatively small diameter.

Within unit 2, the magnetite-ilmenite and magnetite-feldspar ratios tend to increase upward. Within the capping colluvial-soil zone where pyrogenic grains are highly etched, the mineral ratios and feldspar size are larger than elsewhere in the unit.

Local fluctuations in mineral ratios and feldspar sizes are minor in units 3 and 4, but ferromagnesians become increasingly abundant as the rocks lose their red color.

Conclusions.—Initial chemical changes in the lower John Day member pyroclastic debris probably occurred by weathering during deposition (Fisher, 1964). Evidence is that ancient soil zones and *in situ* pisolitic zones occur in the Big Basin section, and that interbedded yellow and red massive tuffs, which grade laterally to lake beds, have apparently retained much of their original character regardless of diagenetic changes that occurred beneath the groundwater table after burial (Hay, 1962, 1963).

The most notable feature of the Big Basin section is the relatively close correlation between mineral alteration and rock colors, especially the concomitant increase of ferromagnesians with loss of red color upward in the section. It could be argued that the ferromagnesians were absent in the original ash low in the section, except that feldspars, magnetite, ilmenite, quartz, apatite, and zircon—associated with them in the upper part of the section—are also present in the lower part.

With present data, it cannot be determined whether the increase in ferromagnesians is owing to a decrease in weathering intensity during deposition, or to an increase in the rate of deposition and resultant rapid burial.

Judging by the similarities in mineral composition throughout the section, and taking into account the absence of ferromagnesians in the lower part of the section because of alteration, the composition of the source magma did not markedly change during deposition. The source locality or localities of the major ash component also probably did not markedly change, as is suggested by the lack of overall vertical trends of (1) feldspar sizes, (2) the magnetite-ilmenite ratio, and (3) the magnetite-feldspar ratio.

Comparison of Lower-Member John Day Rocks with Laterites

CHEMICAL COMPOSITION

Data

The relationships of Al_2O_3, TiO_2, and iron ($Fe_2O_3 + FeO$) are compared here between lower-member John Day rocks and laterites; Fe_2O_3 and FeO are represented by the symbol $\overline{Fe}$. Fresh glass selected to represent the most probable original composition of the altered rocks is from tuff specimens of the upper and middle John Day members. These tuffs have a pyrogenic mineral composition similar to that of most lower-member rocks. Four lower John Day member rocks are compared with specimens of fresh glass. Hawaiian basalts (Macdonald and Katsura,

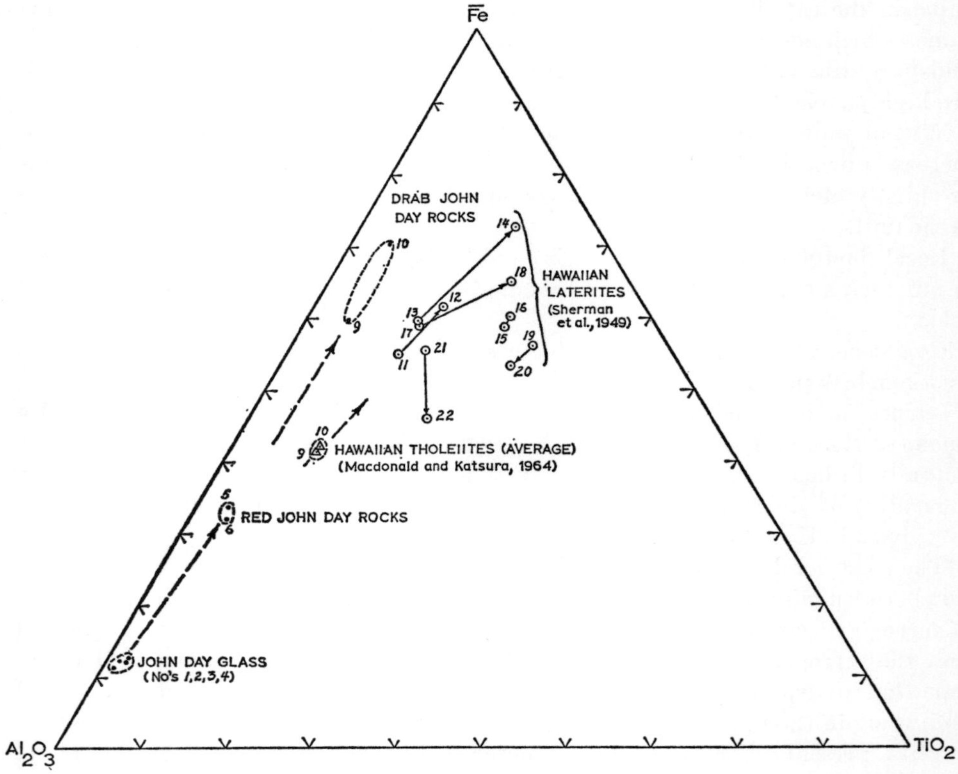

Fig. 4. Ternary diagram of Al_2O_3, TiO_2, and $\overline{Fe}$ for John Day glass, red and drab lower-member John Day rocks, Hawaiian tholeiites, and Hawaiian laterites. $\overline{Fe} = Fe_2O_3 + FeO$. Numbers refer to samples in table 3.

1964) and the A_1 and A_2 horizons from five Hawaiian soil profiles (Sherman *et al.,* 1948) are compared with the John Day rocks. Chemical data for iron, aluminum, and titanium are shown in table 3; a ternary plot for Al_2O_3, $\overline{Fe}$, and TiO_2 is shown in figure 4; TiO_2 versus $[Al_2O_3/(Al_2O_3 + \overline{Fe})]$ is given in figure 5; Al_2O_3 versus $\overline{Fe}$ is given in figure 6; and TiO_2 versus $Al_2O_3 + \overline{Fe}$ is plotted in figure 7.

As suggested by the ternary plot for Al_2O_3, $\overline{Fe}$, and TiO_2 (fig. 4), juvenile John Day ash gained in total iron and TiO_2 relative to Al_2O_3 during alteration to reddish claystone or siltstone, and approaches the composition of Hawaiian tholeiites. The drab John Day rocks contain less TiO_2 and diverge from this trend. Weathering of Hawaiian tholeiites to laterite follows a trend similar to that of the red John Day rocks.

Weathering of the John Day tuffs resulted in a higher rate of iron enrichment than of alumina as TiO_2 increased (fig. 5); lateritic weathering of Hawaiian tholeiites follows about the same trend. In nearly all the Hawaiian laterite soils, the trend continues from the A_1 to the A_2 soil horizon, possibly because Al_2O_3 is leached from the A_2 horizon. The drab lower-member rocks diverge from the trend,

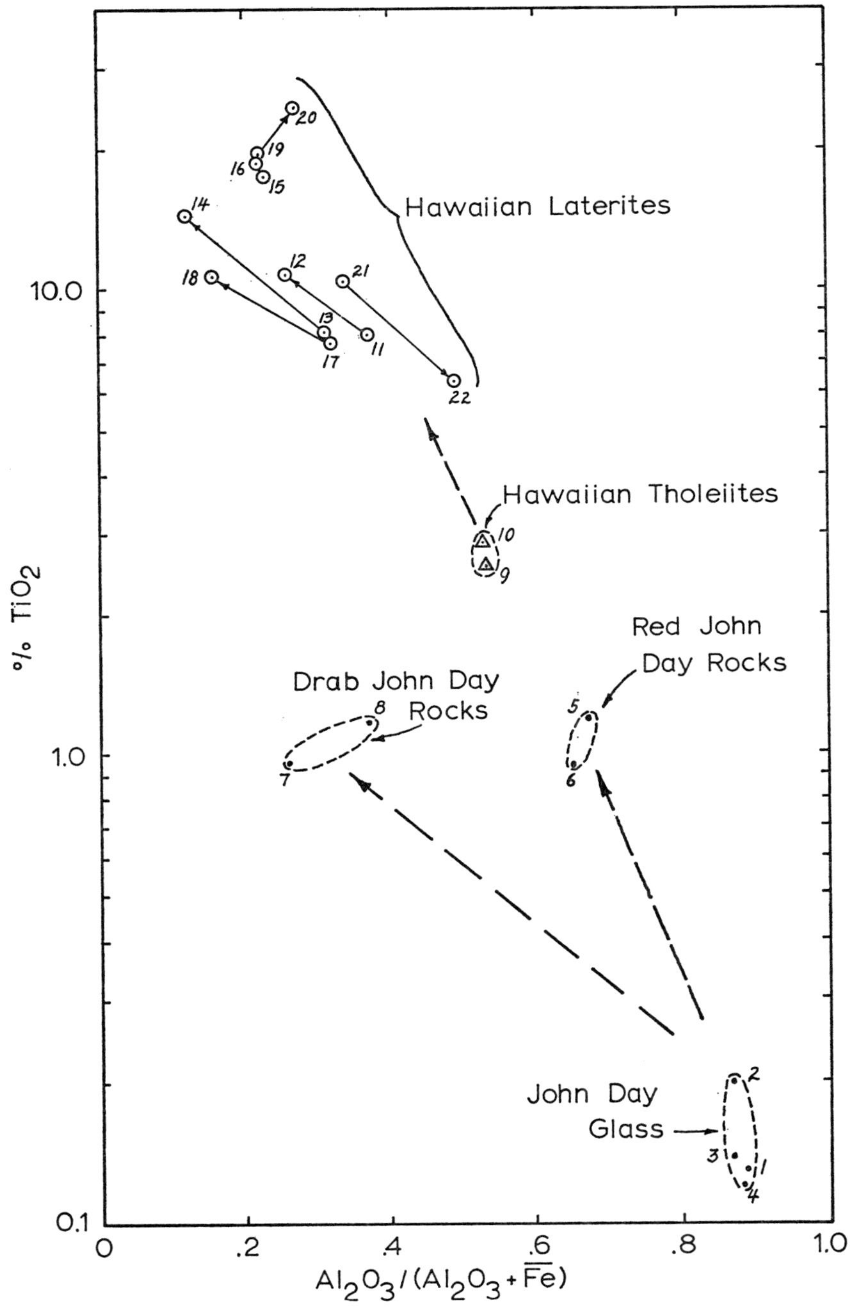

Fig. 5. Aluminum-iron ratio [$Al_2O_3/(Al_2O_3+\overline{Fe})$] versus percentage TiO_2 for John Day glass, red and drab lower-member John Day rocks, Hawaiian tholeiites, and Hawaiian laterites. $\overline{Fe} = Fe_2O_3 + FeO$. Numbers refer to samples in table 3.

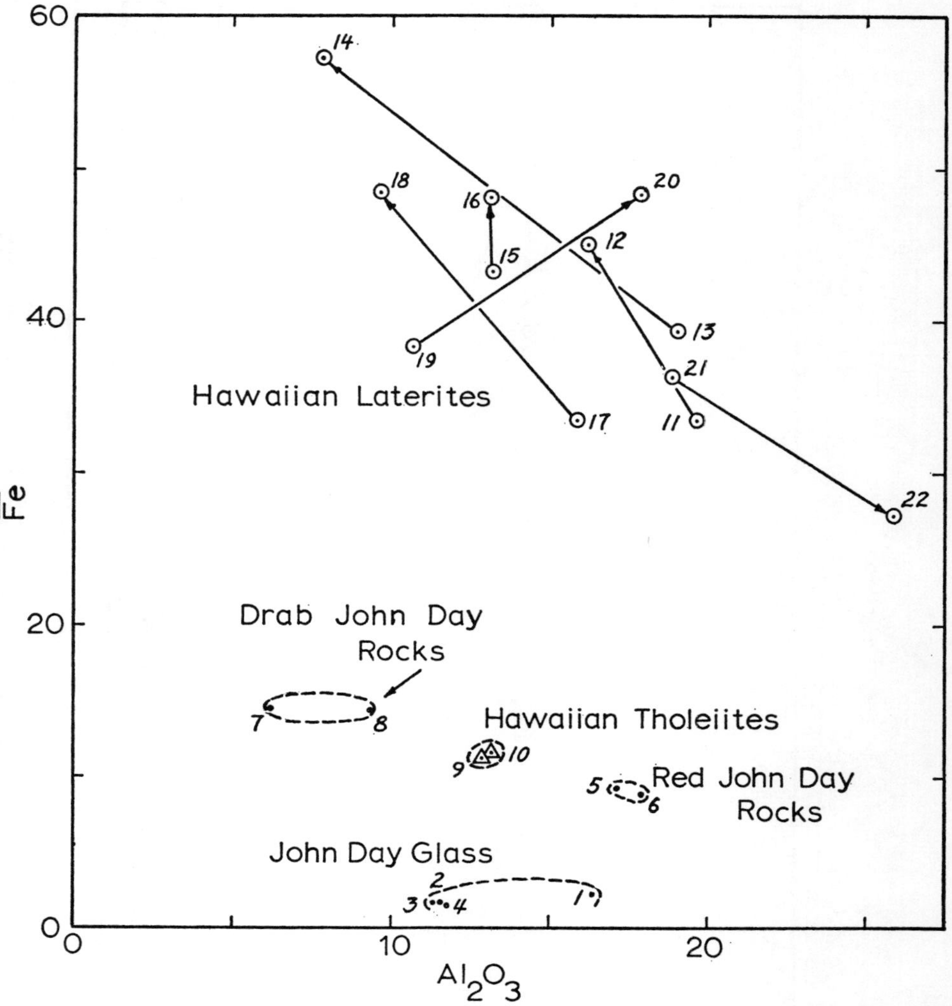

Fig. 6. Al_2O_3 versus $\overline{Fe}$ for John Day glass, red and drab lower-member John Day rocks, Hawaiian tholeiites, and Hawaiian laterites. $\overline{Fe} = Fe_2O_3 + FeO$. Numbers refer to samples in table 3.

because Al_2O_3 is depleted, as is shown in the plot of actual weight percentages of Al_2O_3 and $\overline{Fe}$ (fig. 6).

The plot of actual weight percentages of Al_2O_3 and $\overline{Fe}$ emphasizes the differences of red and drab John Day rocks (fig. 6): the drab rocks contain more iron but less alumina than the red rocks. Most of the A_2 horizons in Hawaiian laterites contain less alumina than the unaltered tholeiites, and all contain relatively more iron; the reason that some contain more alumina is not known. This particular plot shows no apparent continuing trend between the John Day and Hawaiian rocks, but combining Al_2O_3 with $\overline{Fe}$ and plotting them against the weight percentage of TiO_2, as in figure 7, restores the trend shown in previous plots.

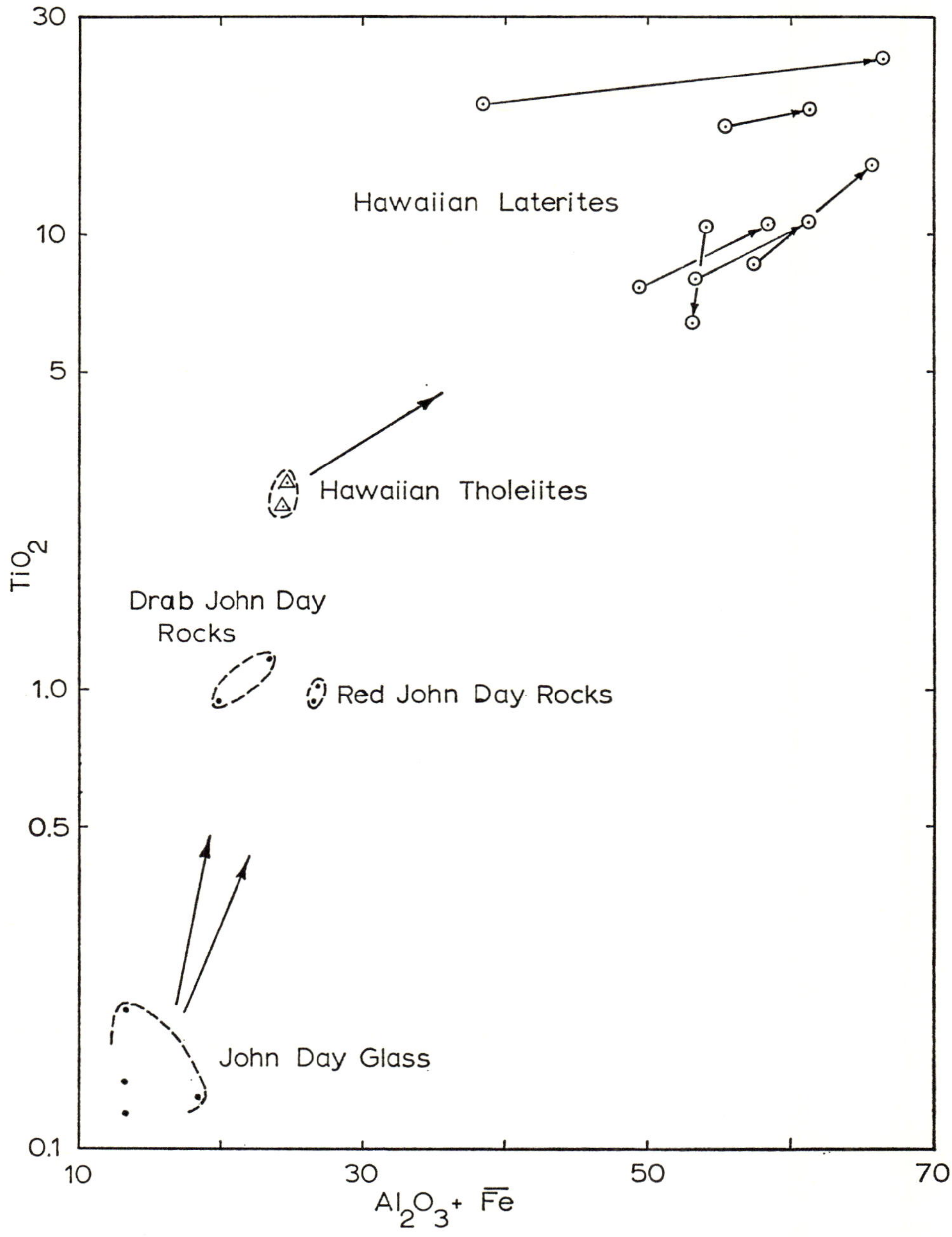

Fig. 7. $Al_2O_3 + \overline{Fe}$ versus TiO_2 for John Day glass, red and drab lower-member John Day rocks, Hawaiian tholeiites, and Hawaiian laterites. $\overline{Fe} = Fe_2O_3 + FeO$. Numbers refer to samples in table 3.

Conclusions

The comparison of Al_2O_3, $\overline{Fe}$, and TiO_2 between Hawaiian laterites and lower-member John Day rocks suggests that the John Day rocks were weathered under conditions similar to those affecting the Hawaiian rocks, but that the process was halted before completion, either by burial, or because weathering conditions were not so severe during lower John Day time.

The chemical divergence of drab and mottled John Day rocks from the red rocks (figs. 4, 5) is probably due to actual leaching of Al_2O_3 from the system Alumina can be mobilized under reducing conditions and within acid soils (Abbott, 1958; Sherman and Ikawa, 1959; Coleman, 1962; Shen and Rich, 1962). Thus the low ratios and the drab color indicate a stage of alteration in a reducing environment, such as a subsurface soil horizon where the groundwater table is high. This conclusion is corroborated by the pyrogenic mineral alteration; most striking are the low percentage of magnetite in the drab rocks, and the fact that the magnetite present is highly pitted.

Laterite Soil Colors and Depositional Environment

The color sequences of rocks in the Big Basin measured section are similar to those of laterite as described in the literature. Laterite soils of Ceylon (Joachim, 1935) are deep red, pinkish red to reddish yellow, and yellowish brown. From Queensland, Australia, Whitehouse (1940, p. 8) describes a laterite sequence, from upper to lower, as deep-red ferruginous layer transitional downward through a mottled zone and then to a light-colored "pallid zone." Where the pallid zone is absent, the mottled zone extends down to the parent rock, and in places the basal zone is silicified. Laterite zones, however, are not necessarily gray, mottled, or iron-rich except where groundwater levels fluctuate (Marbut, 1932, 1934). Laterite developed on andesite ash in Puerto Rico (Catalina clay; Bonnet, 1939) is red to reddish brown without mottled or gray zones.

A typical laterite profile in tropical rain forests, based on a review by Mohr and Van Baren (1954, p. 372), is:

1. Vegetal cover.
2. Light-pink podzolic zone that merges downward to zone 3.
3. A red zone consisting of deep-red clay. The original parent rock structure may be recognizable at the base of the zone.
4. A mottled zone with frequent red, yellow, or violet spots. The original rock structure is visible, especially in the lower parts of the zone.
5. A zone of decomposition which lies just above the parent rock.

In comparison, the color progression of units 1 and 2 in the Big Basin section, with youngest layers first, is:

*Unit 1**	*Unit 2*
1. Red colluvial soil, 5 feet	1. Red colluvial soil, 8 feet
2. Gray, 2 feet	2. Pink, 3 feet
3. Yellow, 3 feet	3. Mottled colluvium, 1 foot
4. Mottled, 2 feet	4. Deep red, 20 feet
5. Deep red, 18 feet	

* Underlain by a hardpan layer that is developed on Cretaceous rocks.

These two lower units in the Big Basin measured section are about the same in thickness and have nearly the same progression of colors. The mottled zones are underlain by deep-red siltstone in each unit rather than by pallid zones as in the Queensland laterite or typical laterite profiles. The red basal zones of each unit lie above well-developed soil structures, and each colluvial zone is underlain by pale-colored and mottled zones. The evidence thus suggests that at one time each colluvial-soil zone formed an upper soil horizon, and that present-day rock colors have not markedly changed since their origin.

Studies by Mohr and Van Baren (1954) and many other workers show that red soils usually develop under oxidizing conditions where soils are well drained; drab colors develop in reducing environments, such as lakes or other places where soil drainage is poor. Fossil leaves from lower-member John Day lake beds (Chaney, 1927) indicate that the climate was tropical or semitropical and that the region was therefore probably heavily vegetated. Lack of vegetal matter in the red beds, therefore, suggests that they formed under oxidizing conditions. Vegetal matter found within the massive drab-colored beds and laminated drab lacustrine beds indicates that they were never highly oxidized and that they have retained their original color. Lack of vegetal matter in many of the drab beds suggests that they may have gone through a state of oxidation.

SUMMARY AND CONCLUSIONS

General Statement

The lower member of the John Day formation was deposited during middle to late Oligocene time. Evidence presented here and elsewhere (Waters, *et al.*, 1951; Hay, 1962; Fisher, 1964) indicates that less than 500 feet of volcanic ash slowly accumulated on a gently rolling, highly weathered erosion surface. Early in John Day time, lowlands were locally dammed and lakes were formed. The topographic surface on which the ash fell was probably in a late mature or early old age of development, and therefore well drained. Thus the development of lakes was unusual. The proximity of basalt flows to lacustrine tuffs in three places in the Picture Gorge quadrangle suggests that the lakes resulted from lava dams.

The John Day rocks of the studied area have had a complex postdepositional history. Each horizon was subjected first to surface weathering and then was gradually covered by later ashfalls. Thus all the particles have been in a variety of chemical environments and accordingly have been altered in varying degree depending upon the length of time spent in each soil environment. The rocks are considered to be a succession of soils which, unlike most soils, did not develop from a static parent; rather, they developed on gradually accumulated volcanic ash which continued to supply fresh material to the chemical budget. Superimposed on the weathering history of the rocks are diagenetic changes that occurred beneath the groundwater table after burial (Hay, 1963).

Volcanic ash responds quickly to changes in its physicochemical environment because of the usually high glass content, high permeability, and abundance of small particles. Resulting secondary mineral assemblages quickly respond to change in the environment; pyrogenic mineral alteration is slower and less sensitive.

Climate During Early John Day Time

Evidence given herein indicates that tuffs of the lower John Day member are lateritic. This implies that the climate was tropical or subtropical, a conclusion consistent with the paleobotanical findings of Chaney (1927).

According to many workers, a semihumid or perhumid tropical climate is essential to form laterite. Mohr and Van Baren (1954) point out, however, that total yearly rainfall is not the only controlling variable. Other factors such as infiltration, runoff and evaporation, organisms, permeability, and relative length of wet and dry seasons may be more important than the total rainfall. Leaching experiments led Pickering (1962) to conclude that progressive desilication proceeds most rapidly in a warm, moist climate where abundant vegetation produces moderately acid groundwater, but that laterites are not necessarily confined to specific climatic regions.

Soil Environment

Initial chemical reactions within the easily altered glass shards differed according to whether they fell on well-drained or poorly drained land, in lakes or swampy lowlands, and according to their rate of deposition. The depositional environments controlled the direction of initial chemical reactions, and the rate of deposition determined how long each particle remained under the influence of a particular soil environment—the direct influence of the atmosphere, the humus layer, and the zones of leaching or accumulation lower in profile.

The presence of kaolinite, montmorillonite, red and drab iron oxides, and organic matter allows qualitative interpretations of soil conditions in initial John Day ash. Diagenetic changes (Hay, 1962, 1963) have been superimposed on changes that occurred in the soil by weathering, but the fact that interbedded red and drab rocks have been preserved indicates that later diagenetic changes have not completely destroyed the secondary minerals produced by weathering.

The presence of kaolinite indicates an acid environment as shown elsewhere by field observations (Mohr and Van Baren, 1954; Gordon and Tracey, 1952). Noll (*in* Marshall, 1964, pp. 79, 112) synthesized kaolin at 300°C in acid systems containing K_2O, Al_2O_3, and SiO_2 in various ratios. Kaolin has also been synthesized at room temperature at pH values from 4.5 to 5.5 in an aqueous solution to which Al and SiO_2 were added (DeKimpe, Gastuche, and Brindley, 1964). As summarized by Keller (1964, p. 17), kaolinite signifies environmental conditions where (1) the ratio of Al to Si is relatively high; (2) H is high, but there is no Na, K, Ca, Mg, or Fe; and (3) the pH is relatively low. These conditions would be met in an acid soil environment where major cations were extensively leached, such as permeable volcanic ash with abundant organic matter, as is visualized for initial John Day ash deposits. The presence of kaolinite together with ferric iron oxides in some of the lower-member John Day rocks indicates an environment with pH less than 7 and a positive Eh.

Montmorillonite forms in alkaline environments with (1) a relatively low Al to Si ratio; (2) relatively abundant Mg, Fe, Ca, Na, and K; and (3) a lower H ion concentration than kaolin (Keller, 1964, p. 22). These conditions would be met in the John Day subsoil where ions could be washed from above. With continuing deposition of ash, early-formed kaolin would be buried, and later would possibly be completely or partly altered to montmorillonite as the environment became increasingly more alkaline and saline. The presence of montmorillonite and ferrous iron oxides in massive drab lower-member John Day beds suggests that the rocks remained below the groundwater table for long periods of time.

The presence of abundant vegetal matter in the original soils doubtless complicated the course of alteration. Until it became depleted, abundant H ions were present in the active soil-forming zone, the zone of intensive biological activity, good aeration, and maximum wetting from atmospheric precipitation.

Mobilization of iron is facilitated by the formation of complex organic molecules with ferrous and ferric hydroxides. $Al(OH)_3$ is more soluble than $Fe(OH)_3$ at low pH values, therefore Fe (and Ti) may become enriched relative to Al. Chemical data from altered lower-member John Day and Hawaiian rocks show this relation.

Lacustrine Tuffs

The lacustrine tuffs, though not lateritic, are composed of fragments deposited contemporaneously with land-laid ash which later became laterized.

In addition to the alteration products montmorillonite, kaolinite, and microcrystalline quartz, and the presence of abundant carbonaceous material in the tuff, gypsum is commonly found in veinlets. Of additional importance to the interpretation of postdepositional environment is the presence of natrojarosite [$NaFe_3(SO_4)_2(OH)_6$] at some localities.

Jarosite [$KFe_3(SO_4)_2(OH)_6$] is a close relative of natrojarosite, and has been more thoroughly studied. Jarosite most commonly occurs in oxidized zones of ore deposits, but also occurs in sedimentary rocks. It occurs within glauconitic sandstones and diatomaceous shales in California (Briggs, 1951), and is associated with illite, kaolinite, and quartz in underclays in Pennsylvania (Warshaw, 1956). Natrojarosite has been described in association with gypsum and carbonized wood from sedimentary rocks (Mitchell and Giannini, 1958).

Insoluble jarosite occurs in association with soluble sulphates of calcium and aluminum. Merwin and Posonjak (1937) report that oxidation of pyrite in the presence of water should yield rhomboclase, $HFe(SO_4)_2 \cdot 4H_2O$, but because the iron oxidizes more slowly than sulphur, mixtures of ferrous and ferric sulphates and sulfuric acid form first. The pH of the solution is raised by the interaction of acid with the rock minerals, and alkalies, if present, are taken into solution. Formation of jarosite and natrojarosite apparently requires oxidizing conditions, and therefore presents the problem of how iron can become mobilized, or how oxidizing conditions can occur underground or in lake beds that contain abundant organic matter.

A possible explanation for the formation of natrojarosite is that plants tend to concentrate sulphur, which in turn may be oxidized by bacterial action in the presence of calcium, and thus may provide sulphate ions. Lyon and Buckman (1946, p. 127) show the following generalized reactions:

Sulphur, combined or elemental,	+	Enzymic oxidation in presence of active Ca	=	SO_3 ions in association with bases, especially Ca,	+	$CO_2 + E$
SO_3 ions	+	Enzymic oxidation in presence of active Ca	=	SO_4 ions in association with bases, especially Ca,	+	$CO_2 + E$

The organisms receive energy from the reaction and give of CO_2. Iron can become mobilized by humic acids and microorganisms in the soil as well as by extracts of fresh leaf litter (Muir *et al.*, 1964).

Natrojarosite is absent in most lacustrine tuffs of the lower John Day member. Why it occurs at just a few localities is not known, but its occurrence in those localities is perhaps related to variations in the concentration of Ca, the abundance of organic matter, and the activity of organisms, which in turn would determine the amount of SO_4, CO_2, Fe, and alkalies in solution.

Pyrogenic Mineral Stability

General Statement

The stability order of pyrogenic minerals in the lower member of the John Day formation differs according to two environments; one order occurs in the red rocks where organic matter is absent, and a second order occurs in drab rocks where organic matter is often present (table 4).

The major differences between the two groupings are: (1) ferromagnesians tend to be absent or highly altered in most of the red rocks; (2) magnetite is more highly etched in the drab than in the red rocks; and (3) the degree of alteration changes among apatite, ilmenite, and magnetite. Clinopyroxene and hornblende occur in drab rocks of the upper Big Basin section, as well as in lacustrine tuff, in pale-pink to gray air-fall tuff, and in massive claystone. Clinopyroxene is less altered in the lacustrine tuff than in the massive drab rocks. Hornblende is less highly etched than clinopyroxene within the same rocks, and does not exhibit the sawtooth terminations typical of altered clinopyroxene.

The relative stability of feldspars is difficult to establish because they are gradational in composition and similar in appearance, but positive identifications made in thin section and from index-of-refraction measurements suggest the stability order shown in table 4.

Discussion

The stability series of pyrogenic minerals in the lower John Day member confirms many of the findings from field and laboratory studies, and may be explained by theoretical considerations of reactivity or mobility of the major cations, and by bonding factors or mineral structures. Mineral stability series have been studied by Goldich (1938), Pettijohn (1941), Graham (1950), Fieldes and Swindale (1954), Jackson and Sherman (1953), and others.

McClelland (1950), in laboratory studies, demonstrated that (1) the release of

TABLE 4

GROUPED STABILITY ORDER OF PYROGENIC MINERALS, LOWER MEMBER JOHN DAY ROCKS

Order	Mineral
	GROUP 1: RED ROCKS, ORGANIC MATTER ABSENT
1......	Zircon, unaltered, sharply euhedral
2......	Magnetite, unpitted to moderately pitted
3......	Ilmenite, slightly pitted to highly etched
4......	Quartz, frosted, stained medium to very pale pink
5......	Apatite, moderately to highly etched with sawtooth terminations; usually absent
6......	Oligoclase, moderately to highly etched, but not raggedly shaped
7......	Andesine, moderately to highly etched, sometimes with ragged shapes
8......	Potassium feldspars, moderately to highly etched and partially altered to clay, usually with cloudy interior
	GROUP 2: DRAB ROCKS, ORGANIC MATTER PRESENT
1......	Zircon, unaltered, sharply euhedral
2......	Apatite, unaltered, euhedral with slightly rounded edges
3......	Ilmenite, unaltered to slightly pitted, sharply euhedral to rounded shapes
4......	Quartz, clear, highly pitted, sharply euhedral to rounded grains
5......	Hornblende, moderately etched, usually anhedral
6......	Magnetite, slightly pitted to highly etched; sometimes absent, or present in trace amounts
7......	Oligoclase, moderately etched
8......	Andesine, moderately etched; etching increases as calcium content increases
9......	Anorthoclase, moderately to highly etched, internally cloudy
10......	Labradorite, moderately to highly etched
11......	Clinopyroxene, highly etched, commonly with sawtooth terminations

TABLE 5

MINERAL STABILITY[a]

Mineral	Mg^{+2}(3.0)[b]	Ca^{+2}(2.0)[b]	Na^{+}(1.0)[b]	K^{+}(0.75)[b]	Total loss (in millimoles/100 gm) of cation after 240 days of treatment
Olivine......................	102	..	..	..	102
Augite......................	17	17	..	..	34
Muscovite......................	...	..	..	31	31
Hornblende......................	16	8	..	..	24
Labradorite......................	...	12	5	..	17
Biotite......................	9	..	..	3	12
Anorthoclase......................	...	..	4	3	7
Microcline......................	...	..	..	5	5
Albite......................	...	..	2	..	2

[a]Modified from McClelland (1950).
[b]Ionic potential in parentheses.

TABLE 6

Stability Parameters of Selected Minerals

Mineral	Reaction energy[a]	Abrasion pH[b]	Bonding energy[c]	Packing index[d]	Bonding index[d]	Linkage factor[d]	Free energy values[e]
Olivine	1.28	10.5	29,796	5.9	0.11	40	(forsterite) −319.8
Ca-Fe Cpx	1.35	10.0	30,728	5.8	0.11	30	....
Ca-Mg Cpx	1.35	10.5	32,052	5.9	0.11	30	....
Hornblende	1.45	10.0	31,883	5.7	0.10	26	....
Anorthite	1.44	8.0	31,935	5.0	0.09	20	....
K-feldspar	1.47	8.0	34,266	4.9	0.08	20	−856.0
Albite	1.49	9.5	34,335	4.9	0.08	20	....
Quartz	1.80	6.5	37,320	4.6	0.08	20	−192.4
Magnetite	1.50	10.0	......	...	....	..	−242.4
Ilmenite	1.40	6.0					
Zircon	1.62	6.0					
Apatite	....	8.0					
John Day glass	....	8.5					
Heulandite	1.67	8.0	......	4.7	0.07	20	....
Hematite	1.50	6.0	......	...	....	..	−177.1
Kaolinite	1.75	6.0	......	...	....	..	−884.5
Montmorillonite	1.77	6.5					
Chlorite	1.58	8.0					
Calcite	1.75	8.0	......	...	....	..	−269.8
Gysum	1.75	6.0	......	...	....	..	−429.2
Jarosite	1.91	5.5					

[a]Reaction energy is calculated by adding electronegativity values with a factor for bonding. Values are from Gruner (1950), except for magnetite, whose values I calculated.

[b]Values are from Stevens and Carron (1948). Abrasion pH represents the ease with which H^+ ions can replace metallic cations, and therefore is related to the mineral structure and the ionic potential of the elements. The higher the value, the greater the ease with which H^+ ions replace the cations.

[c]Values are from Keller (1954).

[d]Values from Poldervaart and Elston (1954).

[e]Free energy values from Garrels and Christ (1965).

bases increases with decreasing particle size, but the influence of particle size varies with different minerals; (2) the rate of base release increases with temperature; (3) the rate of base release decreases with time, suggesting that alteration products around the minerals may retard weathering; (4) the ratio of release of two ions present in the same mineral, with increasing time, tends to approximate the ratio in which these ions are present in the unweathered mineral; (5) the release of bases is apparently accompanied by a breakdown in the crystal lattice; and (6) the order in which bases are released approximates Goldich's (1938) stability series with the exception of muscovite. Table 5, modified from McClelland (1950) and arranged in order of decreasing ionic potential, shows a strong correlation between mineral stability and the ionic potential of the bases. Potassic feldspar is somewhat less stable than sodic feldspar in lower John Day member rocks, the reverse of that inferred by Goldich (1938), but as predicted by McClelland's (1950) data and also by Correns' (1963, table 11, p. 457).

The importance of molecular structure for mineral stability is well known, and can be seen by comparing empirically derived stability series with theoretical parameters such as reaction energy, abrasion pH, bonding energy, packing index, bonding index, and linkage factor (table 6).

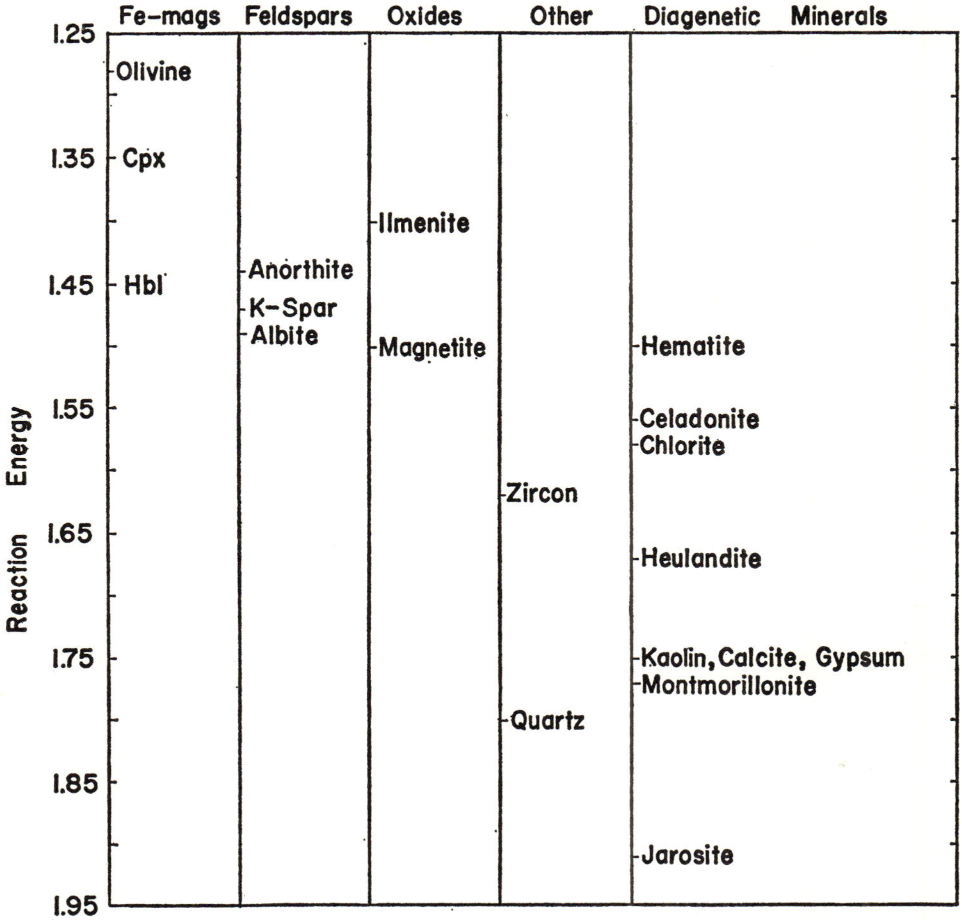

Fig. 8. Reaction energy of minerals in John Day formation. Values are from Gruner (1950).

Reaction energy, calculated by adding electronegativity values with a factor including bond type (Gruner, 1950), shows a mineral stability arrangement similar to that of the lower-member pyrogenic minerals (fig. 8). The values predict potassic feldspar to be somewhat less stable than sodic feldspar, as is observed in the John Day rocks. Ilmenite, with a lower reaction-energy value than magnetite, should alter more rapidly than magnetite as it does in the red tuffs of the lower John Day member. Ilmenite, however, is generally less pitted and predominates over magnetite in many of the drab rocks. Quartz has higher reaction energy than zircon and should be more stable, but quartz is invariably more highly pitted than zircon, and is therefore less stable.

Stevens and Carron (1948) crushed various minerals in distilled water to obtain equilibrium pH values that probably reflect exchange of H ions for metallic cations. Some of their data are arranged in sequence in figure 9. Minerals with high pH values are generally stable in solutions of high pH and are most easily altered in a low pH environment. Thus, according to abrasion pH, magnetite

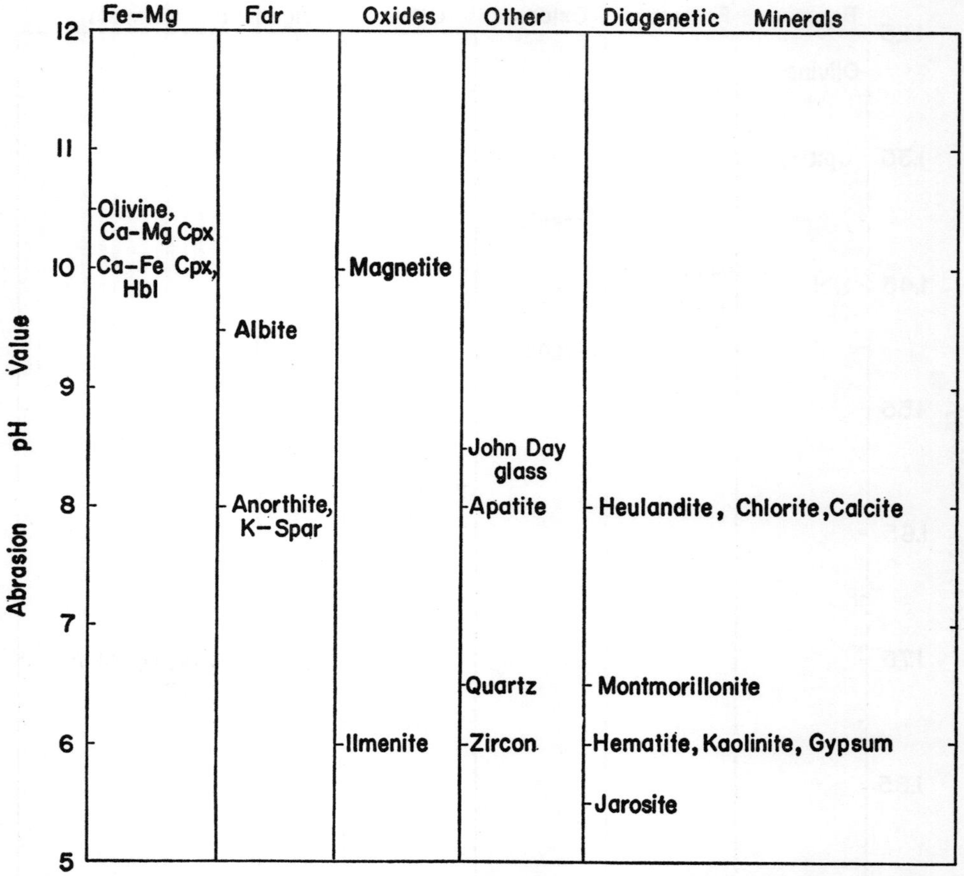

Fig. 9. Abrasion pH of minerals in the lower member of the John Day formation. Values are from Stevens and Carron (1948); value of John Day glass is from Hay (1963).

should be less stable than ilmenite, quartz less stable than zircon, and albite less stable than potassium feldspar. Quartz should be most stable, and olivine, clinopyroxene, and magnetite least stable, in an acid environment. Comparison with lower-member John Day pyrogenic minerals shows that stability is not predicted as closely by abrasion pH as by the reaction energy.

Bonding energies, determined by Keller (1954) for some silicates, increase in value with an increase in stability. Packing index, bonding index, and linkage factor (Poldervaart and Elston, 1954), all parameters of mineral structure, decrease in value with increase in stability. The packing index is defined as ten times the volume ratio of the unit cell divided into the total volume of ions in the unit cell (Fairbairn, 1943); the bonding index is the number of bonds per cubic angstrom, obtained by dividing the volume of the unit cell into the unit cell; the linkage factor is as follows:

$$\frac{\text{total O not in OH}^-}{\text{Si}^{+4} + (\text{Al}^{+3} \text{ in 4-coordination}) \times 10}$$

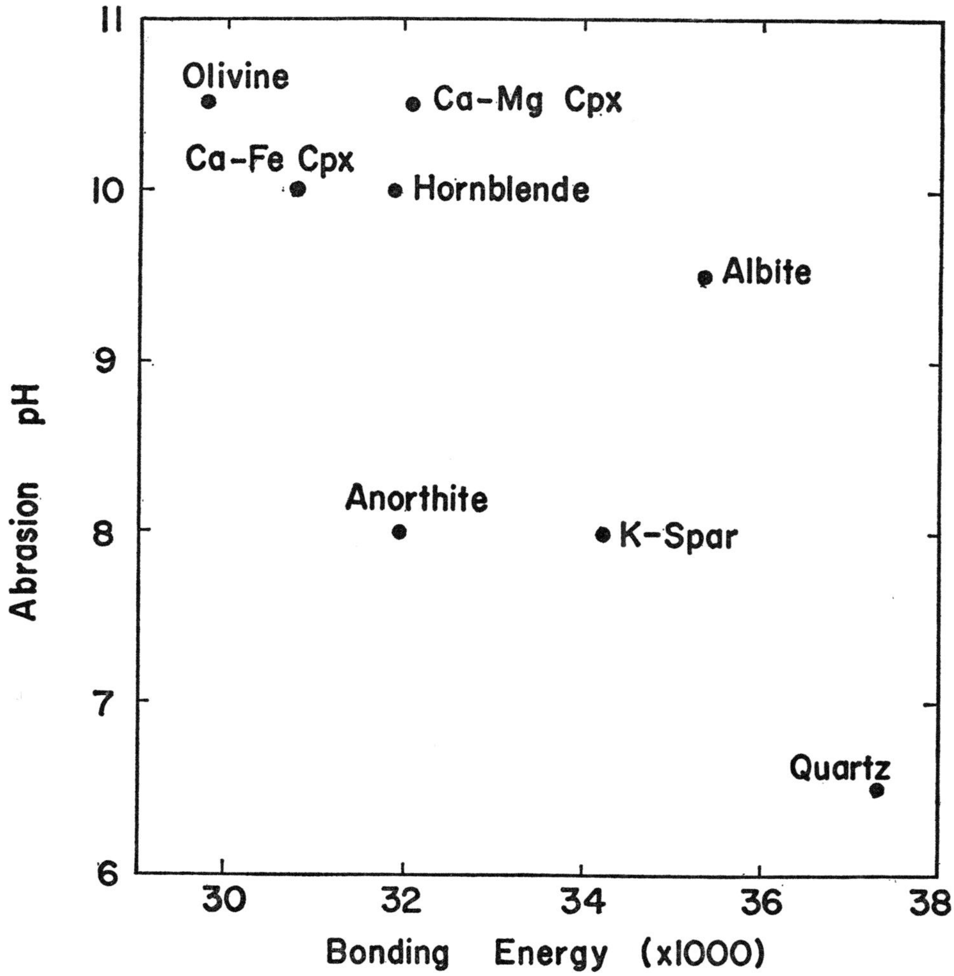

Fig. 10. Bonding energy (Keller, 1954) of rock-forming silicates plotted against abrasion pH (Stevens and Carron, 1948).

Relationships between some of the structural parameters are plotted in figures 10, 11, and 12. Figure 10 shows that abrasion pH tends to increase as bonding energy decreases, and is a rough measure of the influence of bonding on the mobility of major cations.

Bonding energy versus reaction energy (fig. 11) shows a systematic arrangement of the pyrogenic minerals found in rocks of the lower John Day member. Although bond type is included in calculations of reaction energy (Gruner, 1950), bonding-energy values are derived from the energy of formation of cation-oxygen bonds from a reference gaseous state (Keller, 1954). The plot suggests that molecular structure and ionic potentials have a nearly equal influence on mineral stability.

Packing index versus reaction energy (fig. 12) shows that as the number of ions in a unit cell decreases relative to cell size (i.e., the value for packing index de-

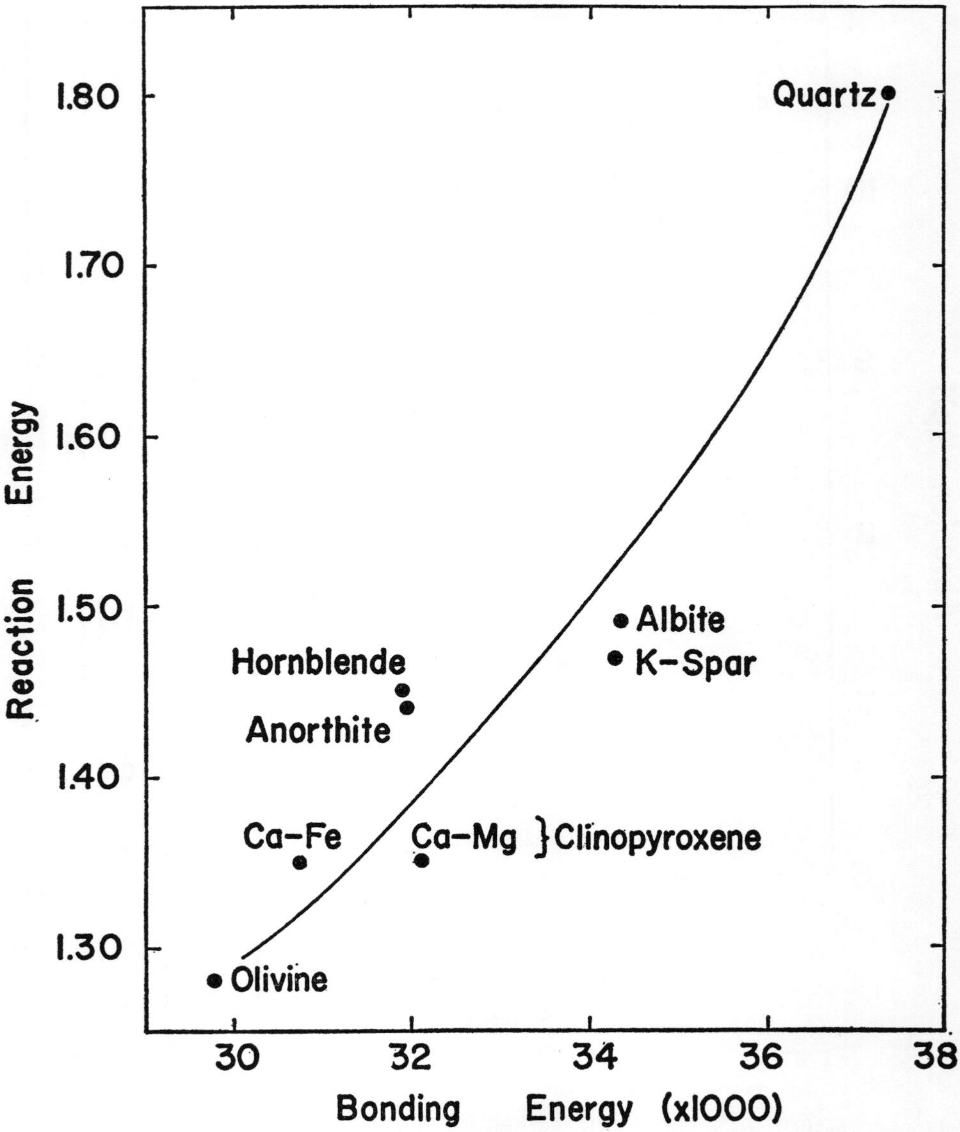

Fig. 11. Bonding energy (Keller, 1954) of rock-forming silicates plotted against reaction energy (Gruner, 1950).

creases), the reaction energy increases. This scatterplot most effectively separates the feldspars and ferromagnesian minerals and closely follows the general stability order observed in lower-member John Day rocks. Thus it appears that these two parameters most effectively predict the weathering stability of the pyrogenic minerals.

Inferred Conditions of Eh and pH

Stratigraphic evidence indicates that many of the red and drab lower-member John Day rocks are laterally equivalent; differences in secondary and pyrogenic

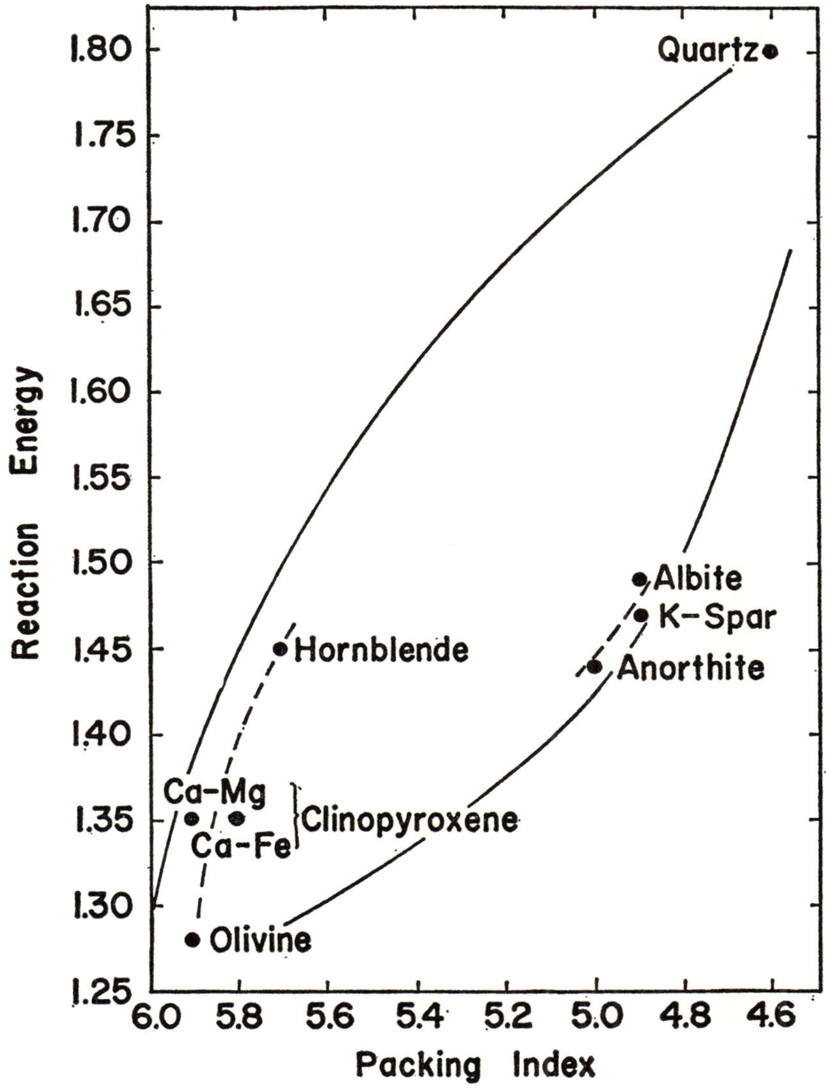

Fig. 12. Reaction energy (Gruner, 1950) of rock-forming silicates plotted against packing index (Poldervaart and Elston, 1954).

minerals of the red and drab rocks suggest that changes in mineralogy were caused by different chemical conditions controlled by somewhat different environments. These environments are qualitatively illustrated by an Eh-pH diagram (fig. 13) as set forth by Krumbein and Garrels (1952) and Garrels and Christ (1965, p. 381). The fields A, B, and C and their boundaries are tentative and purposely left vague. The sequence of pyrogenic mineral stability for field A is given in group 1 (table 4); for fields B and C they are given in group 2.

Field A represents rocks containing ferric oxide, montmorillonite, and kaolinite,

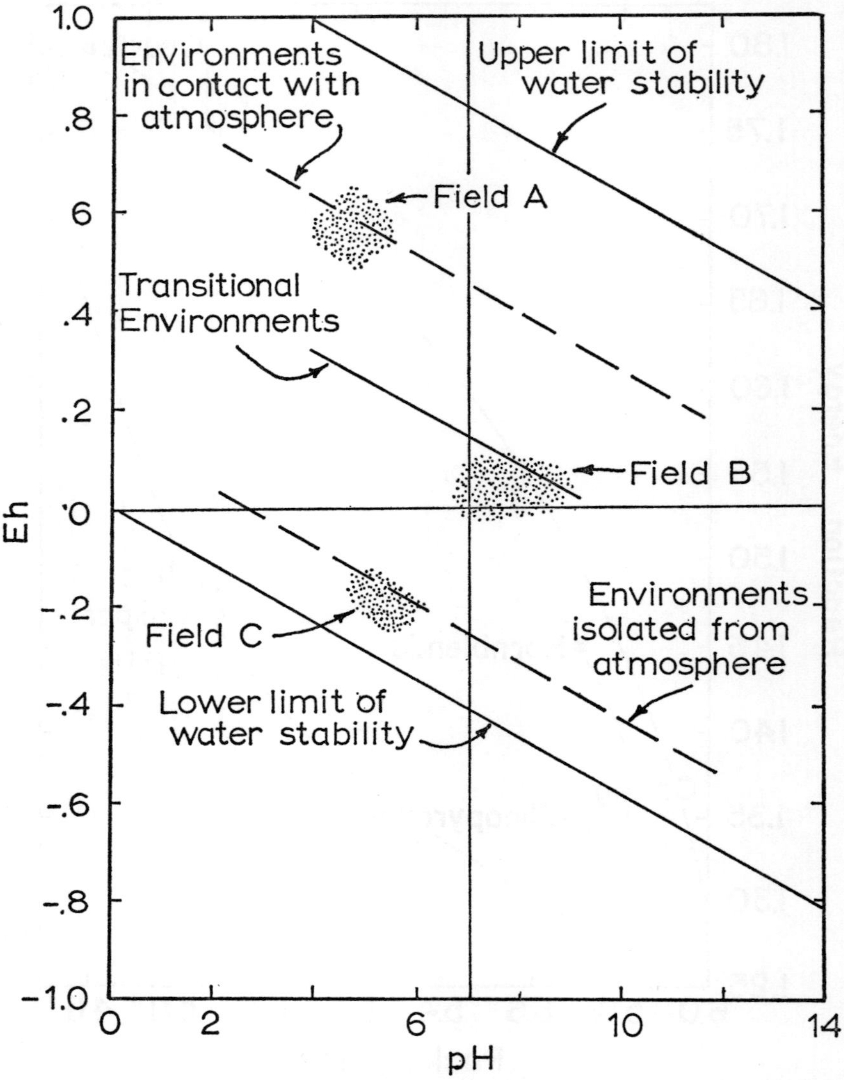

Fig. 13. Inferred conditions of Eh and pH for lower-member John Day pyrogenic minerals based on diagram as set forth by Krumbein and Garrels (1952) and Garrels and Christ (1965). Fields A, B, and C are defined in text.

and includes the deep-red lateritic lower-member tuffs where relict shard and pumice textures are often absent.

Field B is in the clinoptilolite facies–montmorillonite subfacies of Hay (1963). The presence of montmorillonite suggests alkaline pH. Some beds contain small amounts of organic material, but it is usually absent. This suggests a stability field that overlaps the boundary between a positive and a negative Eh environment, although many of the rocks may have passed through an initial oxidizing environment that destroyed original organic matter.

Field C is defined by the presence of organic matter and kaolin, and the absence of ferric oxides. Fixation of gypsum and natrojarosite in some of the lacustrine rocks at a later time indicates that the Eh-pH field did not markedly change after burial. The pyrogenic mineral stability within the lacustrine tuffs is essentially the same as in field B, but in most of the beds, pyrogenic minerals are sparse, and many of those in group 2 are absent in some specimens. Quartz is rare, and highly pitted magnetite forms a trace or is absent.

Diagenetic minerals in the altered tops of olivine basalt (calcite, celadonite, chlorite, and chalcedony) indicate a low Eh environment; cogenetic calcite indicates pH values higher than in the lacustrine or massive drab beds.

Extensive silicification of rocks in the lower member of the John Day formation is rare, and occurs most commonly beneath the lateritic red rocks. The lacustrine tuff and altered olivine basalt are locally silicified, and in each, silica was precipitated last in sequence. It was introduced after natrojarosite in the shale and later than iron oxide in the basalt.

Hay (1963) deals primarily with diagenetic relations among montmorillonite, clinoptilolite, and celadonite, which, except for celadonite, occur within the red lower-member John Day claystones and siltstones. Clinoptilolite replaces shards and pumice in some of the red lower-member rocks, particularly in the upper part of the Big Basin section. Its presence with ferric iron oxides is difficult to explain because it probably forms under conditions of relatively high activity ratio of (Na + K) to H (Hay, 1963, p. 237), conditions that are reversed during lateritic weathering. Apparently the ferric oxides remained stable after burial, even with a decrease in Eh and an increase in pH.

LITERATURE CITED

ABBOTT, A. T.
1958. Occurrence of gibbsite on the island of Kauai, Hawaii Islands. Econ. Geol., vol. 53, pp. 842–853.

BONNET, J. A.
1939. The nature of laterization as revealed by chemical, physical and mineralogical studies of a lateritic soil profile from Puerto Rico. Soil Sci., vol. 48, pp. 25–40.

BRIGGS, L. I., JR.
1951. Jarosite from the California Tertiary. Am. Min., vol. 36, pp. 902–906.

CHANEY, R. W.
1927. Fossil floras of the John Day Basin. Carnegie Inst. Washington Publ. 346, pp. 45–138.

COLEMAN, N. T.
1962. Decomposition of clays and the fate of aluminum. Econ. Geol., Vol. 57, pp. 1207–1218.

CORRENS, C. W.
1963. Experiments on the decomposition of silicates and discussion of chemical weathering. *In* Clays and clay minerals. 10th Conf. New York: Pergamon Press. Pp. 443–459.

DEKIMPE, C., M. C. GASTUCHE, and G. W. BRINDLEY
1964. Low temperature synthesis of kaolin minerals. Am. Min., vol. 49, pp. 1–16.

FAIRBAIRN, H. W.
1943. Packing in ionic minerals. Geol. Soc. Amer Bull., vol. 54, pp. 1305–1374.

FIELDES, M., and L. D. SWINDALE
1954. Chemical weathering of silicates in soil formation. New Zealand Jour. Sci. Technol., vol. 36B, pp. 140–154.

FISHER, R. V.
1963. Bubble-wall texture and its significance. Jour. Sed. Petrol., vol. 3, pp. 224–227.
1964. Resurrected Oligocene hills, eastern Oregon. Am. Jour. Sci., vol. 262, pp. 713–725.
1966. The geology of a Miocene ignimbrite layer, John Day formation, eastern Oregon. Univ. Calif. Publ. Geol. Sci., vol. 67. 73 pp.
1967. Early Tertiary deformation in north-central Oregon. Amer. Assoc. Petrol. Geol., vol. 51, pp. 111–123.

FISHER, R. V., and R. E. WILCOX
1960. The John Day formation in the Monument quadrangle, Oregon. U.S. Geol. Survey Prof. Paper 400-B, pp. 302–304.

GARRELS, R. M., and C. L. CHRIST
1965. Solutions, minerals and equilibria. New York: Harper and Row. 450 pp.

GOLDICH, S. S.
1938. A study in rock-weathering. Jour. Geol., vol. 46, pp. 17–58.

GORDON, M., JR., and J. I. TRACEY, JR.
1952. Origin of the Arkansas bauxite deposits. *In* Clay and laterite genesis. Symposium, Am. Inst. Min. Met. Eng. New York. pp. 12–34.

GRAHAM, E. R.
1950. The plagioclase feldspars as an index to soil weathering. Soil Sci. Soc. Amer., Proc., vol. 14, pp. 300–302.

GRUNER, J. W.
1950. An attempt to arrange silicates in the order of their reaction energies at relatively low temperatures. Am. Min., vol. 35, pp. 137–148.

HAY, R. L.
1962. Origin and diagenetic alteration of the lower part of the John Day formation near Mitchell, Oregon. Geol. Soc. Amer., Buddington Volume, pp. 191–216.
1963. Stratigraphy and zeolitic diagenesis of the John Day formation of Oregon. Univ. Calif. Publ. Geol. Sci., vol. 42, no. 5, pp. 199–262.

JACKSON, M. L., and G. D. SHERMAN
1953. Chemical weathering of minerals in soils. Advances in Agron., vol. 5. New York: Academic Press. Pp. 219–318.

JOACHIM, A. W. R.
1935. Studies on Ceylon soils, II: General characteristics of Ceylon soils, some typical soil groups of the island, and a tentative scheme of classification. Trop. Agriculturist (Ceylon), vol. 84, pp. 254–274.
JONES, D. L.
1960. Lower Cretaceous (Albian) fossils from southwestern Oregon and their paleogeographic significance. Jour. Paleontol., vol. 34, pp. 152–160.
KELLER, W. D.
1954. Bonding energies of some silicate minerals. Am. Min., vol. 39, pp. 783–793.
1964. Processes of origin and alteration of clay minerals. *In* Soil clay mineralogy, ed. C. I. Rich and G. W. Kunze. Chapel Hill: University of North Carolina Press. Pp. 3–76.
KRUMBEIN, W. C.
1934. Size frequency and distribution of sediments. Jour. Sed. Petrol., vol. 4, pp. 65–77.
KRUMBEIN, W. C., and R. M. GARRELS
1952. Origin and classification of chemical sediments in terms of pH and oxidation-reduction potentials. Jour. Geol., vol. 60, pp. 1–33.
LINDSLEY, D. H.
1960. Geology of the Spray quadrangle, Oregon. Unpublished Ph.D. thesis, Johns Hopkins University. 236 pp.
LYON, T. L., and H. O. BUCKMAN
1946. The nature and properties of soils. New York: Macmillan. 499 pp.
MCCLELLAND, J. E.
1950. The effect of time, temperature, and particle size on the release of bases from some common soil-forming minerals of different crystal structure. Soil Sci. Soc. Amer., Proc., vol. 15, pp. 301–307.
MACDONALD, G. A., and T. KATSURA
1964. Chemical composition of Hawaiian lavas. Jour. Petrol., vol. 5, pp. 82–133.
MARBUT, C. F.
1932. Morphology of laterites. Proc. 2d Intl. Cong. Soil Sci., 1930, vol. 5, pp. 72–80.
1934. The work of Commission V of the International Society of Soil Science. Soil Res., vol. 4, pp. 139–147.
MARSHALL, C. E.
1964. The physical chemistry and mineralogy of soils, vol. 1: Soil materials. New York: Wiley and Sons. 388 pp.
MERRIAM, J. C.
1901. A contribution to the geology of the John Day Basin. Univ. Calif. Publ., Bull. Dept. Geol. Sci., vol. 2, pp. 269–314.
MERWIN, H. E., and E. POSONJAK
1937. Sulphate incrustations in the Copper Queen Mine, Bisbee, Arizona. Am. Min., vol. 22, pp. 567–571.
MITCHELL, R. S., and W. F. GIANNINI
1958. Natrojarosite from near the Montana-Wyoming line. Am. Min., vol. 43, pp. 1205–1210.
MOHR, E. C. J., and F. A. VAN BAREN
1954. Tropical soils. New York: Interscience Publishers. 498 pp.
MUIR, J. W., R. I. MORRISON, C. J. BROWN, and J. LOGAN
1964. The mobilization of iron by aqueous extracts of plants, I. Jour. Soil Sci., vol. 15, pp. 220–237.
PACKARD, E. L.
1956. An *Engonoceras* from central Oregon. Jour. Paleontol., vol. 30, pp. 398–402.
PECK, D. L.
1964. Geological reconnaisance of the Antelope-Ashwood area, north-central Oregon. U.S. Geol. Survey Bull. 1161-D. 26 pp.
PETTIJOHN, F. J.
1941. Persistence of heavy minerals and geologic age. Jour. Geol., vol. 49, pp. 610–625.
PICKERING, R. J.
1962. Some leaching experiments on three quartz-free silicate rocks and their contribution to an understanding of laterization. Econ. Geol., vol. 57, pp. 1185–1206.

POLDERVAART, A., and W. E. ELSTON

1954. The calc-alkaline series and the trend of fractional crystallization of basaltic magma: a new approach to graphic representation. Jour. Geol., vol. 62, pp. 150–162.

SHEN, M. J., and C. I. RICH

1962. Aluminum fixation in montmorillonite. Soil. Sci. Soc. Amer., Proc., vol. 26, pp. 33–36.

SHERMAN, G. D., Z. C. FOSTER, and C. K. FUJIMOTO

1948. Some of the properties of the ferruginous humic latosols of the Hawaiian Islands. Soil Sci. Soc. Amer., Proc., vol. 13, pp. 471–476.

SHERMAN, G. D., and H. IKAWA

1959. Occurrence of gibbsite amygdules in Haiku bauxite area of Maui. Pacific Sci., vol. 13, pp. 291–294.

SMITH, J. R.

1958. The optical properties of heated plagioclases. Am. Min., vol. 43, pp. 1179–1194.

STEVENS, R. E., and M. K. CARRON

1948. Simple field test for distinguishing minerals by abrasion pH. Am. Min., vol. 33, pp. 31–49.

WARSHAW, C. M.

1956. The occurrence of jarosite in underclays. Am. Min., vol. 41, pp. 288–296.

WATERS, A. C.

1961. Stratigraphic and lithologic variations in the Columbia River basalt. Am. Jour. Sci., vol. 259, pp. 583–611.

WATERS, A. C., R. E. BROWN, R. R. COMPTON, L. W. STAPLES, G. W. WALKER, and H. WILLIAMS

1951. Quicksilver deposits of the Horse Heaven mining district, Oregon. U.S. Geol. Survey Bull. 969-E. 149 pp.

WHITEHOUSE, F. W.

1940. Studies in the late geological history of Queensland, I: The lateritic soils of western Queensland. Univ. Queensland Pap. Dept. Geol., n.s., vol. 2, no. 1, pp. 2–22.

PLATE

PLATE 1

Fig. 1. Oligoclase crystal surrounded by rim of montmorillonite; lower border of crystal shows alteration to clay along its length (×350).

Fig. 2. Sodic labradorite with highly etched borders and internal darkened areas from alteration in a lateritic tuff (×400).

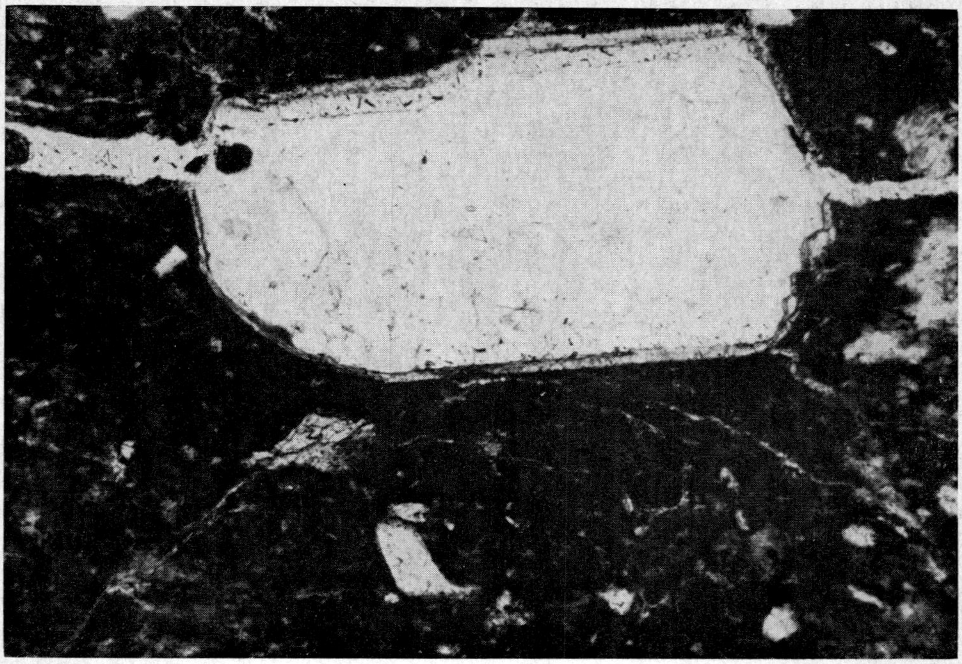

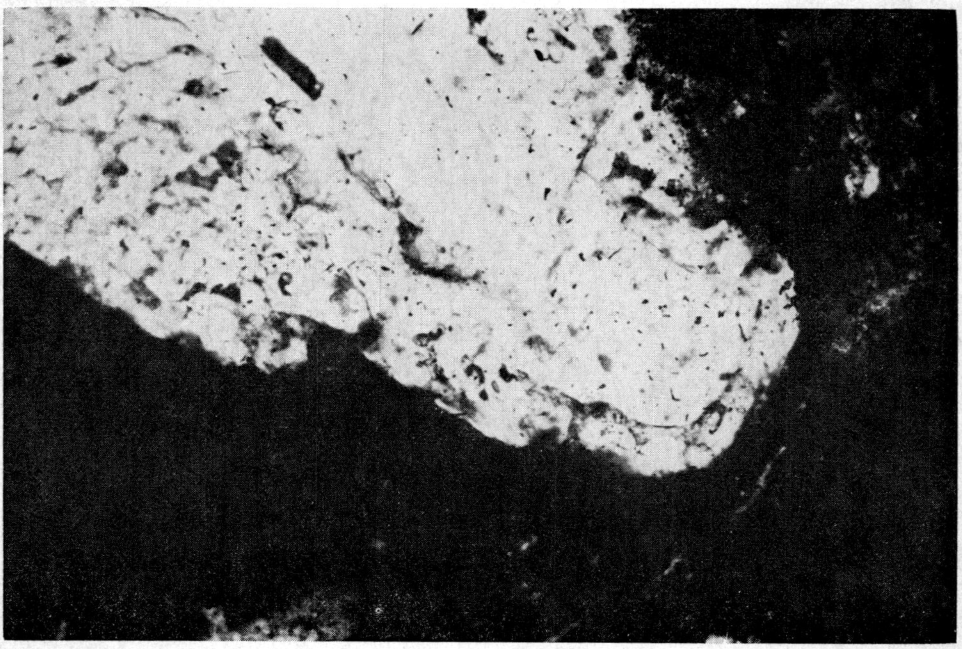